Engineering Treatment of Soils

OTHER TITLES FROM E & FN SPON

Cyclic Loading of Soils
Edited by M.P. O'Reilly and S.F. Brown

Drainage Design
Edited by P. Smart and J.G. Herbertson

Durability of Geotextiles
RILEM

Foundations on Rock
D.C. Wyllie

Geology of Construction Materials
J.E. Prentice

Geology for Civil Engineers
A. McLean and C. Gribble

Geomembranes – Identification and Performance Testing
Edited by A.L. Rollin and J.M. Rigo

Ground Improvement
Edited by M.P. Mosely

Ground Subsidence
A.C. Waltham

Introducing Groundwater
M. Price

Pile Design and Construction Practice
M.J. Tomlinson

Rock Mechanics for Underground Mining
B.H.G. Brady and E.T. Brown

Soil Mechanics
R.F. Craig

Structural Foundations Manual for Low-Rise Buildings
M.F. Atkinson

The Stability of Slopes
E.N. Bromhead

Underpinning and Retention
Edited by S. Thorburn and G.S. Littlejohn

For information on these and other titles please contact:
The Promotion Department, E & FN Spon, 2–6 Boundary Row,
London SE1 8HN. Telephone: 071-865 0066

Engineering Treatment of Soils

F.G. Bell

Department of Geology and Applied Geology,
University of Natal, Durban

E & FN SPON
An Imprint of Chapman & Hall

London · Glasgow · New York · Tokyo · Melbourne · Madras

Published by E & FN Spon, an imprint of Chapman & Hall, 2–6 Boundary Row, London SE1 8HN

Chapman & Hall, 2–6 Boundary Row, London SE1 8HN, UK

Blackie Academic & Professional, Wester Cleddens Road, Bishopbriggs, Glasgow G64 2NZ, UK

Chapman & Hall Inc., 29 West 35th Street, New York NY10001, USA

Chapman & Hall Japan, Thomson Publishing Japan, Hirakawacho Nemoto Building, 6F, 1-7-11 Hirakawa-cho, Chiyoda-ku, Tokyo 102, Japan

Chapman & Hall Australia, Thomas Nelson Australia, 102 Dodds Street, South Melbourne, Victoria 3205, Australia

Chapman & Hall India, R. Seshadri, 32 Second Main Road, CIT East, Madras 600 035, India

First edition 1993

© 1993 F.G. Bell

Typeset in 10/12 Palatino by Keyboard Services, Luton

Printed in Great Britain by
St. Edmundsbury Press, Bury St. Edmunds, Suffolk

ISBN 0 419 17750 7

A catalogue record for this book is available from the British Library

Library of Congress Cataloging-in-Publication data

Bell, F.G. (Frederic Gladstone)
    Engineering treatment of soils / F.G. Bell. -- 1st ed.
        p.   cm.
    Includes bibliographical references and index.
    ISBN 0-419-17750-7 (alk. paper)
    1. Soil mechanics.   I. Title.
TA710.B425   1993
621.1'5136--dc20
                                                                92-40044
                                                                CIP

Contents

Preface

The engineering treatment of soils involves improving their geotechnical character for construction purposes. Problems regarding construction in soils arise from their lack of strength which manifests itself in their deformation, which beneath foundations takes the form of settlement, or in some exceptional circumstances gives rise to ground failure. Problems are also associated with water in soil. Hence soil treatment is primarily concerned with enhancing its strength or reducing or excluding the water content. The techniques used to improve soil conditions may be temporary or permanent. For example, groundwater lowering techniques and freezing are generally temporary in nature whilst grouting is permanent. Also some techniques have a long history whilst others have evolved during the last 25 years, notably some of the methods of soil reinforcement, and dynamic compaction. Moreover new techniques are continually being developed.

Some of the techniques used to treat soils are highly specialized and as such may not be very familiar to many of those involved with construction, notably civil and structural engineers and engineering geologists, and to a lesser extent builders and architects. Accordingly this text attempts to provide those in, or intending to enter, the construction industry with a basic appreciation of the various techniques concerned with soil treatment, under which circumstances a particular technique will be used, and in which soil types a given method can be used successfully.

Obviously the depth to which the subject can be dealt with is limited by the space available (in fact the book is about one third larger than the publishers initially agreed to). Consequently no attempt has been made to cover detailed design methods of the various techniques mentioned. Some design methods, such as those used in soil reinforcement are dealt with in specialist textbooks, others such as some of those used in ground freezing may have been developed by, and so are peculiar to, the specialist contractors concerned. However, for those who are interested in pursuing the subject to greater depth there is a list of suggested further readings, together with an extensive list of references. The reader will see from the references that some journals and proceedings such as Ground Engineering; Proceedings of the American Society of Civil Engineers – Journal of the Geotechnical Engineering Division; and the Proceedings of the International Conferences on Soil Mechanics and Foundation Engineering are mentioned frequently. These, together with specialist journals or conference proceedings such as the Journal of Geotextiles and Geomembranes or the international conferences on Ground Freezing and Geotextiles, provide particularly worthwhile sources to keep one up to date.

F.G. Bell, 1992

1

Introduction

Most civil engineering operations are carried out in soil and, obviously, poor soil conditions will be encountered on some construction sites. If such soil cannot be removed, then its engineering behaviour can often be enhanced by some method of ground treatment (Glossop, 1968). Poor soil conditions usually are attributable to an excess of groundwater or a lack of strength, and associated deformability. Treatment methods are therefore aimed at preventing ingress of groundwater to or removing it from the site in question on the one hand or improving soil strength on the other.

Groundwater flow becomes more significant as the permeability of the soil increases. The fact that groundwater flow is much slower in silts and clays than in sands and gravels does not mean that there are no problems associated with groundwater in the two former soil types. For example, quick conditions and piping are associated with silts and ground heave with expanding clay soils. Conversely, inundation with water can lead to some loess soils collapsing. Other problems associated with groundwater are dissolution of minerals such as gypsum in gypsiferous sands and the hydration of others, for example, anhydrite. Yet other minerals break down rapidly in the presence of groundwater, like pyrite; the resulting sulphate ions, when carried in solution, can attack concrete foundations.

Soils with low strength are also highly deformable. Lack of strength leads to soil failing if it is overloaded. However, this is not a frequent occurrence in civil engineering construction. Much more important is soil deformation which, in terms of structures erected at the ground surface, gives rise to settlement. Some of the most problematic soils include: peat and organic soils; quick clays, residual montmorillonitic clays and varved clays, which may be sensitive to extra-sensitive; and loosely packed saturated alluvial, estuarine or marine sands, silts and muds.

Soil treatment techniques may be either temporary or permanent. For example, the use of a wellpoint system for dewatering can be regarded as a temporary technique, as can freezing. On the other hand, grouting is a permanent method of ground treatment. The type of technique chosen depends on the nature of the problem and the type of soil conditions. Cost is obviously a factor that enters into the equation.

Soil is one of the most abundant and cheapest of construction materials. Even so its use can be greatly extended by enhancing its engineering performance, for example, by the addition of cementitious material or by incorporation of reinforcing elements. The concept of soil reinforcement was developed over 20 years ago and is now commonly employed in soil structures, notably in embankments and retaining walls. In this context, the use of geosynthetics has expanded enormously but they can be used for other purposes such as filtration and separation.

1.1 SITE INVESTIGATION

Any important ground improvement works must be preceded by a site investigation to establish the type and succession of soils that occur at the site concerned. The methods of investigation necessitated by a programme of ground improvement involve those methods which are generally used in soil investigation (Clayton *et al.*, 1982; Weltman and Head, 1983). Obviously an evaluation and selection of the most suitable improvement technique can only be made after a thorough picture of the ground conditions is established. Some of the investigation techniques used prior to the ground improvement operation should be able to be used during and after the improvement works so that the effectiveness of the work can be assessed.

In some cases an investigation may need to be more extensive when the same technique is being used, but for different purposes. For example, investigations prior to soils being grouted will probably need to be more extensive when this method is being used to form a cutoff curtain for a dam than when it is used to enhance the ground for foundation purposes.

In addition to establishing the geology of the site, the investigation might also have to unravel the history of the site. This is especially the case in urban areas where the former use of the site may be important. For example, there may be obstructions, voids, waste materials etc., present which may adversely affect the use of certain improvement measures. Full-scale testing on site should be carried out if the applicability of the method in particular conditions cannot be confirmed by routine tests.

Samples are required for conventional laboratory testing to ascertain the properties of the soils concerned. This not only aids the selection of the treatment process but also is required for the design of the ground improvement programme. For instance, data on consolidation are required when the soil is to be improved by preloading.

Ground improvement works frequently need monitoring while they are being carried out. Again taking preloading, the amount of settlement needs to be monitored during the operation. Similarly, pore pressures in clayey soils need to be monitored between tamping runs during dynamic compaction.

Then when the ground investigation programme has been completed, the effectiveness of the treatment may need to be assessed. This is not the case if the method of ground improvement employed is a temporary method. Various methods can be employed from taking samples and testing to see if the strength has increased or if the permeability has been reduced, to carrying out some field test (e.g. permeability testing after grouting; pressuremeter tests for dynamic compaction; standard or cone penetration tests after vibroflotation). Seismic methods have also been used to assess ground improvement including crosshole seismic, crosshole radar and interborehole acoustic emission. A survey of some of these methods and recent developments has been provided by Rathmeyer and Saari (1983).

1.2 SOIL CLASSIFICATION

Casagrande (1948) advanced one of the first comprehensive engineering classifications of soil. In the Casagrande system the coarse-grained soils are distinguished from the fine on a basis of particle size. Gravels and sands are the two principal types of coarse-grained soils and in this classification both are subdivided into five subgroups on a basis of grading (Table 1.1). Each of the main soil types and subgroups are given a letter, a pair of which are combined in the group symbol, the former being the prefix, the latter the suffix. Fine-grained soils are subdivided on a basis of their plasticity (Table 1.1). Subsequently the Unified Soil Classification (Table 1.2) was developed from the Casagrande system.

The British Soil Classification for engineering purposes (Anon., 1981(b)) also uses particle size as a fundamental parameter. Classification can be made either by rapid assessment in the field or by full laboratory procedure (Tables 1.3 and 1.4 respectively). Boulders (B; over 200 mm), cobbles (Cb; 60–200 mm), gravels (G; 2–60 mm), sands (S; 0.06–2 mm), silts (M; 0.002–0.06 mm) and clays (C; less than 0.002 mm) are distinguished as individual groups. Mixed soil types are given in Figure 1.1.

These major soil groups are again divided into subgroups on a basis of

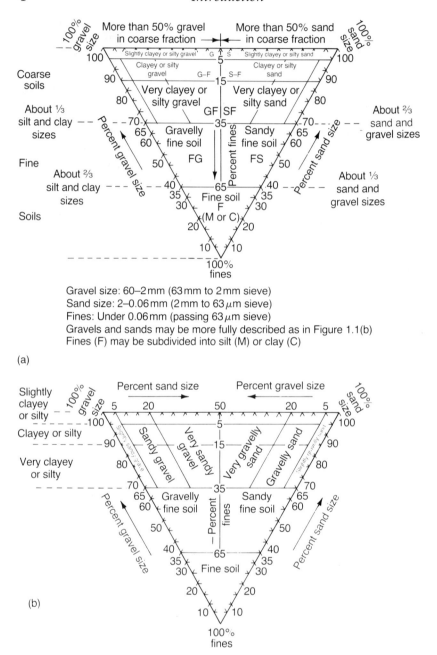

Gravel size: 60–2 mm (63 mm to 2 mm sieve)
Sand size: 2–0.06 mm (2 mm to 63 μm sieve)
Fines: Under 0.06 mm (passing 63 μm sieve)
Gravels and sands may be more fully described as in Figure 1.1(b)
Fines (F) may be subdivided into silt (M) or clay (C)

(a)

(b)

Figure 1.1 (a) Grading triangle for soil classification (material finer than 60 mm); (b) Fuller description of gravels and sands. (Reproduced by permission of the Chief Executive, Transport Research Laboratory, Crowthorne, Berkshire)

Table 1.1 Symbols used in the Casagrande soil classification

Main soil type		*Prefix*
Coarse-grained soils	Gravel	G
	Sand	S
Fine-grained soils	Silt	M
	Clay	C
	Organic silts and clays	O
Fibrous soils	Peat	Pt
Subdivisions		*Suffix*
For coarse-grained soils	Well graded, with little or no fines	W
	Well graded with suitable clay binder	C
	Uniformly graded with little or no fines	U
	Poorly graded with little or no fines	P
	Poorly graded with appreciable fines or well graded with excess fines	F
For fine-grained soils	Low compressibility (plasticity)	L
	Medium compressibility (plasticity)	I
	High compressibility (plasticity)	H

grading in the case of cohesionless soils, and on a basis of plasticity in the case of fine material. Granular soils are described as well graded (W) or poorly graded (P). Two further types of poorly graded granular soils are recognized, namely, uniformly graded (Pu) and gap-graded (Pg). Silts and clays are generally subdivided according to their liquid limits (LL) into low (under 35%), intermediate (35–50%), high (50–70%), very high (70–90%) and extremely high (over 90%) subgroups. As in the Casagrande classification, each subgroup is given a combined symbol in which the letter describing the predominant size fraction is written first (e.g. GW = well-graded gravels; CH = clay with high liquid limit; see Table 1.4).

Any group may be referred to as organic if it contains a significant proportion of organic matter, in which case the letter O is suffixed to the group symbol (e.g. CVSO = organic clay of very high liquid limit with sand). The symbol Pt is given to peat.

In many soil classifications boulders and cobbles are removed before an attempt is made at classification; for example, their proportions are recorded separately in the British Soil Classification. Their presence should be recorded in the soil description, a plus sign being used in symbols for soil mixtures, for example, G + Cb for gravel with cobbles. The British Soil Classification groups very coarse deposits as follows:

1. **Boulders** Over half of the very coarse material is of boulder size (over 200 mm); may be described as cobbly boulders if cobbles are an important second constituent in the very coarse fraction.

Table 1.2 Unified Soil Classification (after Wagner, 1957)

Field identification procedures (excluding particles larger than 76 mm and basing fractions on estimated weights)				Group symbols*	Typical names	Information required for describing soils	Laboratory classification criteria
Coarse-grained soils More than half of material is larger than No. 200 sieve size‡	Gravels More than half of coarse fraction is larger than No. 7 sieve‡	Clean gravels (little or no fines)	Wide range in grain size and substantial amounts of all intermediate particle sizes	GW	Well graded gravels, gravel-sand mixtures, little or no fine	Give typical name; indicate approximate percentages of sand and gravel; maximum size; angularity, surface condition, and hardness of the coarse grains; local or geologic name and other pertinent descriptive information: and symbols in parentheses	$C_u = \dfrac{D_{60}}{D_{10}}$ Greater than 4 $C_c = \dfrac{(D_{30})^2}{D_{10} \times D_{60}}$ Between 1 and 3
			Predominantly one size or a range of sizes with some intermediate sizes missing	GP	Poorly graded gravels, gravel-sand mixtures, little or no fine		Not meeting all gradation requirements for GW
		Gravels with fines (appreciable amount of fines)	Nonplastic fines (for identification procedures see ML below)	GM	Silty gravels, poorly graded gravel-sand-silt mixtures	For undisturbed soils add information on stratification, degree of compactness, cementation, moisture conditions and drainage characteristics	Atterberg limits below 'A' line, or PI less than 4 / Atterberg limits above 'A' line with PI greater than 7 — Above 'A' line with PI between 4 and 7 are borderline cases requiring use of dual symbols
			Plastic fines (for identification procedures, see CL below)	GC	Clayey gravels, poorly graded gravel-sand-clay mixtures		
	Sands More than half of coarse fraction is smaller than No. 7 sieve size‡	Clean sands (little or no fines)	Wide range in grain sizes and substantial amounts of all intermediate particle sizes	SW	Well graded sands, gravelly sands, little or no fines	Example: Silty sand, gravelly; about 20% hard, angular gravel particles 12.5 mm maximum size; rounded and subangular sand grains coarse to fine, about 15% non-plastic fines with low dry strength; well compacted and moist in place; alluvial sand: (SM)	$C_u = \dfrac{D_{60}}{D_{10}}$ Greater than 6 $C_c = \dfrac{(D_{30})^2}{D_{10} \times D_{60}}$ Between 1 and 3
			Predominantly one size or a range of sizes with some intermediate sizes missing	SP	Poorly graded sands, gravelly sands, little or no fines		Not meeting all gradation requirements for SW
		Sands with fines (appreciable amount of fines)	Nonplastic fines (for identification procedures see ML below)	SM	Silty sands, poorly graded sand-silt mixtures		Atterberg limits below 'A' line, or PI less than 5 / Atterberg limits above 'A' line with PI greater than 7 — Above 'A' line with PI between 4 and 7 are borderline cases requiring use of dual symbols
			Plastic fines (for identification procedures, see CL below)	SC	Clayey sands, poorly graded sand-clay mixtures		

Determine percentages of gravel and sand from grain size curve. Depending on fines (fraction smaller than No. 200 sieve size) coarse-grained soils are classified as follows: Less than 5% GW, GP, SW, SP. More than 12%: GM, GC, SM, SC. 5% to 12%: Borderline cases require use of dual symbols

Use grain size curve in identifying the fractions as given under field identification

* Boundary classifications. Soils possessing characteristics of two groups are designated by combinations of group symbols. For example GW–GC, well-graded gravel-sand mixture with clay binder.
† All sieve sizes on this chart are US standard.
‡ For visual classification, the 6.3 mm size may be used as equivalent to the No. 7 sieve size.
Field identification procedure for fine-grained soils or fractions
These procedures are to be performed on the minus No. 40 sieve size particles, approximately 0.4 mm. For field classification purposes, screening is not intended, simply remove by hand the coarse particles that interfere with the tests.

Plasticity chart

Plasticity index (PI): 0, 10, 20, 30, 40, 50, 60

Comparing soils at equal liquid limit
Toughness and dry strength increase with increasing plasticity index

Zones: CL-ML, ML, CL, OL or ML, CH, OH or MH

A-line

Liquid limit: 0 10 20 30 40 50 60 70 80 90 100

Plasticity chart for laboratory classification of fine-grained soils

Use grain size curve in identifying the functions as given under field identification

Fine-grained soils More than half of material is smaller than No. 200 sieve size†	Identification procedures on fraction smaller than No. 40 sieve size				
	DRY STRENGTH (crushing characteristics)	DILATANCY (reaction to shaking)	TOUGHNESS (consistency near plastic limit)		
smaller than No. 200 sieve size†					Give typical name: indicate degree and character of plasticity, amount and maximum size of coarse grains; colour in wet condition, odour if any, local or geologic name, and other pertinent descriptive information, and symbol in parentheses
Silts and clays liquid limit less than 50	None to slight	Quick to slow	None	ML	Inorganic silts and very fine sands, rock flour, silty or clayey fine sands with slight plasticity
	Medium to high	None to very slow	Medium	CL	Inorganic clays of low to medium plasticity, gravelly clays, sandy clays, silty clays, lean clays
	Slight to medium	Slow	Slight	OL	Organic silts and organic silt-clays of low plasticity
					For undisturbed soils add information on structure, stratification, consistency in undisturbed and remoulded states, moisture and drainage conditions
Silts and clays liquid limit greater than 50	Slight to medium	Slow to none	Slight to medium	MH	Inorganic silts micaceous or diatomaceous fine sandy or silty soils, elastic silts
	High to very high	None	High	CH	Inorganic clays of high plasticity, fat clays
					Example
	Medium to high	None to very slow	Slight to medium	OH	Organic clays of medium to high plasticity; *Clayey silt, brown; slightly plastic; small percentage of fine sand; numerous vertical root holes; firm and dry in place; loess; (ML)*
Highly organic soils	Readily identified by colour, odour, spongy feel and frequently by fibrous texture			Pt	Peat and other highly organic soils

Dilatancy (reacting to shaking) After removing particles larger than No. 40 sieve size, prepare a pat of moist soil with a volume of about 1 cm³. Add enough water if necessary to make the soil soft but not sticky. Place the pat in the open palm of one hand and shake horizontally, striking vigorously against the other hand several times. A positive reaction consists of the appearance of water on the surface of the pat which changes to a livery consistency and becomes glossy. When the sample is squeezed between the fingers, the water and gloss disappear from the surface, the pat stiffens and finally it cracks and crumbles. The rapidity of appearance of water during shaking and of its disappearance during squeezing assisting in identifying the character of the fines in a soil. Very fine clean sands give the quickest and most distinct reaction whereas a plastic clay has no reaction. Inorganic silts, such as a typical rock flour, show a moderately quick reaction.

Dry strength (crushing characteristics) After removing particles larger than No. 40 sieve size, mould a pat of soil to the consistency of putty, adding water if necessary. Allow the pat to dry completely by oven, sun or air drying, and then test its strength by breaking and crumbling between the fingers. This strength is a measure of the character and quantity of the colloidal fraction contained in the soil. The dry strength increases with increasing plasticity. High dry strength is characteristic for clays of the CH group. A typical inorganic silt possesses only very slight dry strength. Silty fine sands and silts have about the same slight dry strength, but can be distinguished by the feel when powdering the dried specimen. Fine sand feels gritty whereas a typical silt has the smooth feel of flour.

Toughness (consistency near plastic limit) After removing particles larger than the No. 40 sieve size, a specimen of soil about 1 cm³ in size, is moulded to the consistency of putty. If too dry, water must be added and if sticky, the specimen should be spread out in a thin layer and allowed to lose some moisture by evaporation. Then the specimen is rolled out by hand on a smooth surface or between the palms into a thread about 3 mm in diameter. The thread is then folded and re-rolled repeatedly. During this manipulation the moisture content is gradually reduced and the specimen stiffens, finally loses its plasticity, and crumbles when the plastic limit is reached. After the thread crumbles, the pieces should be lumped together and a slight kneading action continued until the lump crumbles. The tougher the thread near the plastic limit and the stiffer the lump when it finally crumbles, the more potent is the colloidal clay fraction in the soil. Weakness of the thread at the plastic limit and quick loss of coherence of the lump below the plastic limit indicate either inorganic clay of low plasticity, or materials such as kaolin-type clays and organic clays which occur below the A-line. Highly organic clays have a very weak and spongy feel at the plastic limit.

Table 1.3 Field identification and description of soils (after Anon, 1981(b))

	Basic soil type	Particle size (mm)		Visual identification	Particle nature and plasticity	Composite soil types (mixtures of basic soil types)	
Very coarse soils	**Boulders**			Only seen complete in pits or exposures.	Particle shape:	*Scale of secondary constituents with coarse soils*	
	Cobbles	200 — 60		Often difficult to recover from boreholes	Angular Subangular	Term	% of clay or silt
Coarse soils (over 65% sand and gravel sizes)	**Gravels**	coarse	20	Easily visible to naked eye; particle shape can be described; grading can be described.	Subrounded Rounded Flat Elongate	Slightly clayey **Gravel** or **Sand** Slightly silty	under 5
		medium	6	Well graded: wide range of grain sizes, well distributed. Poorly graded: not well graded. (May be uniform: size of most particles lies between narrow limits; or gap graded: an intermediate size of particle is markedly under-represented.)		— clayey **Gravel** or — silty **Sand**	5 to 15
		fine	2		Texture:	Very clayey **Gravel** or **Very silty** **Sand**	15 to 35
	Sands	coarse	0.6	Visible to naked eye; very little or no cohesion when dry; grading can be described.	Rough Smooth Polished	Sandy **Gravel** Gravelly **Sand**	Sand or gravel as important second constituent of the coarse fraction
		medium	0.2	Well graded: wide range of grain sizes, well distributed. Poorly graded: not well graded. (May be uniform: size of most particles lies between narrow limits; or gap graded: an intermediate size of particle is markedly under-represented.)		For composite types described as: clayey: fines are plastic, cohesive: silty: fines non-plastic or of low plasicity	
		fine	0.06				
Fine soils (over 35% silt and clay sizes)	**Silts**	coarse	0.02	Only coarse silt barely visible to naked eye; exhibits little plasticity and marked dilatancy; slightly granular or silky to the touch. Disintegrates in water; lumps dry quickly; possess cohesion but can be powdered easily between fingers.	Non-plastic or low plasticity	*Scale of secondary constituents with fine soils*	
		medium	0.006			Term	% of sand gravel
		fine	0.002			sandy **Clay** or gravelly **Silt**	35 to 65
	Clays			Dry lumps can be broken but not powered between the fingers; they also disintegrate under water but more slowly than silt; smooth to the touch; exhibits plasticity but no dilatancy; sticks to the fingers and dries slowly; shrinks appreciably on drying usually showing cracks. Intermediate and high plasticity clays show these properties to a moderate and high degree, respectively.	Intermediate plasticity (Lean clay)	— CLAY:SILT	Under 35
						Examples of composite types	
					High plasticity (Fat clay)	(Indicating preferred order for description)	
						Loose, brown, subangular very sandy, fine to coarse **gravel** with small pockets of soft grey clay	
Organic soils	**Organic clay, silt or Sand**	Varies		Contains substantial amounts of organic vegetable matter.		Medium dense, light brown, clayey, fine and medium **sand**	
	Peats	Varies		Predominantly plant remains usually dark brown or black in colour, often with distinctive smell; low bulk density.		Stiff, orange brown, fissured sandy **clay** Firm, brown, thinly laminated **silt** and **clay** Plastic, brown, amorphous **Peat**	

Compactness/strength		Structure			Colour
Term	Field test	Term	Field identification	Interval scales	
Loose	By inspection of voids and particle packing.	Homogeneous	Deposit consists essentially of one type	*Scale of bedding spacing*	Red Pink Yellow Brown Olive Green Blue White Grey Black, etc.
Dense		Inter-stratified	Alternating layers of varying types or with band or lenses of other materials. Interval scale for bedding spacing may be used.	Term / Mean spacing (mm)	
				Very thickly bedded — Over 2000	
				Thickly bedded — 2000-600	
Loose	Can be excavated with a spade; 50 mm wooden peg can be easily driven.	Hetero-geneous	A mixture of types	Medium bedded — 600-200	
				Thinly bedded — 200-60	
Dense	Requires pick for excavation; 50 mm wooden peg hard to drive.	Weathered	Particles may be weakened and may show concentric layering.	Very thinly bedded — 60-20	Supplemented as necessary with:
				Thickly laminated — 20-6	
Slightly cemented	Visual examination; pick removes soil in lumps which can be abraded.			Thinly laminated — under 6	Light Dark Mottled, etc.
					and
Soft or loose	Easily moulded or crushed in the fingers				
		Fissured	Break into polyhedral fragments along fissures. Interval scale for spacing of discontinuities may be used.		Pinkish Reddish Yellowish Brownish, etc.
Firm or dense	Can be moulded or crushed by strong pressure in the fingers.				
Very soft	Exudes between fingers when squeezed in hand.	Intact	No fissures.		
Soft	Moulded by light finger pressure.	Homogeneous	Deposit consists essentially of one type.	*Scale of spacing of other discontinuities*	
Firm	Can be moulded by strong finger pressure.	Inter-stratified	Altering layers of varying types. Interval scale for thickness of layers may be used. Usually has crumb or columnar structure.	Term / Mean spacing (mm)	
Stiff	Cannot be moulded by fingers. Can be indented by thumb.	Weathered		Very widely spaced — Over 2000	
Very stiff	Can be indented by thumb nail.			Widely spaced — 2000-600	
Firm	Fibres already compressed together			Medium spaced — 600-200	
				Closely spaced — 200-60	
Spongy	Very compressible and open structure.	Fibrous	Plant remains recognisable and retain some strength.	Very closely spaced — 60-20	
Plastic	Can be moulded in hand, and smears fingers.	Amor-phous	Recognisable plant remains absent.	Extremely closely spaced — Under 30	

Table 1.4 British Soil Classification System for Engineering Purposes (after Anon. 1981(b))

Soil groups (see note 1) GRAVEL and SAND may be qualified Sandy GRAVEL and Gravelly SAND, etc. where appropriate	Subgroups and laboratory identification					
		Group Symbol (see notes 2 & 3)	Subgroup symbol (see note 2)		Fines (% less than 0.06 mm)	Liquid limit (%)	Name
COARSE SOILS less than 35% of the material is finer than 0.06 mm — **GRAVELS** More than 50% of coarse material is of gravel size (coarser than 2 mm)	Slightly silty or clayey GRAVEL	G	GW GP	GW GPu GPg	0 to 5		Well graded GRAVEL Poorly graded/Uniform/Gap graded GRAVEL
	Silty GRAVEL	G-F	G-M	GWM GPM	5 to 15		Well graded/Poorly graded silty GRAVEL
	Clayey GRAVEL		G-C	GWC GPC			Well graded/Poorly graded clayey GRAVEL
	Very silty GRAVEL	GF	GM	GML, etc.	15 to 35		Very silty GRAVEL) subdivided as for GC
	Very clayey GRAVEL		GC	GCL GCI CCH GCV GCE			Very clayey GRAVEL (clay of low, intermediate, high, very high, extremely high plasticity)
SANDS More than 50% of coarse material is of sand size (finer than 2 mm)	Slightly silty or clayey SAND	S	SW SP	SW SPu SPg	0 to 5		Well graded SAND Poorly graded/Uniform/Gap graded SAND
	Silty SAND	S-F	S-M	SWM SPM	5 to 15		Well graded/Poorly graded silty SAND
	Clayey SAND		S-C	SWC SPC			Well graded/Poorly graded clayey SAND
	Very silty SAND	SF	SM	SML, etc.	15 to 35		Very silty SAND; subdivided as for SC
	Very clayey SAND		SC	SCL SCI SCH SCV SCE			Very clayey SAND (clay of low, intermediate, high, very high, extremely high plasticity)

Table 1.4 cont.

FINE SOILS (more than 35% of the material is finer than 0.06 mm)	Soil name			Group / subgroup symbols			%	Description
Gravelly or sandy SILTS and CLAYS: 35% to 65% fines	Gravelly SILT	FG		MG		MLG, etc		Gravelly SILT; subdivide as for CG
	Gravelly CLAY (see note 4)			CG		CLG	<35	Gravelly CLAY of low plasticity
						CIG	35 to 50	of intermediate plasticity
						CHG	50 to 70	of high plasticity
						CVG	70 to 90	of very high plasticity
						CEG	>90	of extremely high plasticity
	Sandy SILT (see note 4)	FS		MS		MLS, etc.		Sandy SILT; subdivide as for CG
	Sandy CLAY			CS		CLS, etc.		Sandy CLAY; subdivide as for CG
SILTS and CLAYS: 65% to 100% fines	SILT (M-SOIL)	F		M		ML, etc.		SILT; subdivide as for C
	CLAY (see notes 5 & 6)			C		CL	<35	CLAY of low plasticity
						CI	35 to 50	of intermediate plasticity
						CH	50 to 70	of high plasticity
						CV	70 to 90	of very high plasticity
						CE	>90	of extremely high plasticity
ORGANIC SOILS	Descriptive letter 'O' suffixed to any group or subgroup symbol.							Organic matter suspected to be a significant constituent. Example MHO: Organic SILT of high plasticity.
PEAT	Pt							Peat soils consist predominantly of plant remains which may be fibrous or amorphous.

Notes:

1. The name of the soil group should always be given when describing soils, supplemented, if required, by the group symbol, although for some additional applications (e.g. longitudinal sections) it may be convenient to use the group symbol alone.
2. The group symbol or subgroup symbol should be placed in brackets if laboratory methods have not been used for identification, e.g. (GC).
3. The designation FINE SOIL, or FINES, F, may be used in place of SILT, M, or CLAY, C, when it is not possible or not required to distinguish between them.
4. GRAVELLY if more than 50% of coarse material is of gravel size. SANDY if more than 50% of coarse material is of sand size.
5. SILT (M-SOIL), M, is material plotting below the 'A' line, and has a restricted plastic range in relation to its liquid limit, and relatively low cohesion. Fine soils of this type include clean silt-sized materials and rock flour, micaceous soils, pumice, and volcanic soils containing halloysite. The alternative term 'M-soil' avoids confusion with materials of predominantly silt size, which form only a part of the group. Organic soils also usually plot below the 'A' line on the plasticity chart, when they are designated ORGANIC SILT, MO.
6. CLAY, C, is material plotting above the 'A' line, and is fully plastic in relation to its liquid limit.

Table 1.5 Description of material/soil mixtures

Term	Composition
Boulders (or **Cobbles**) with a little finer material*	Up to 5% finer material
Boulders (or **Cobbles**) with some finer material*	5–20% finer material
Boulders (or **Cobbles**) with much finer material*	20–50% finer material
Finer material* with many **Boulders** (or **Cobbles**)	50–20% boulders (or cobbles)
Finer material* with some **Boulders** (or **Cobbles**)	20–5% boulders (or cobbles)
Finer material* with occasional **Boulders** (or **Cobbles**)	Up to 5% boulders (or cobbles)

* Give the name of the finer material (in parentheses when it is is the minor constituent), e.g. cobbly **boulders** with some finer material (sand with some fines).

2. **Cobbles**   Over half of the very coarse material is of cobble size (200–60 mm); may be described as bouldery cobbles if boulders are an important second constituent in the very coarse fraction.

Mixtures of very coarse material and soil can be described by combining the terms for the very coarse constituent and the soil constituent, as shown in Table 1.5.

1.3 SOIL TYPES

1.3.1 Coarse-grained soils

Size and sorting have a significant influence on the engineering behaviour of granular soils. For example, the void ratios of well-sorted and perfectly cohesionless aggregates of equidimensional grains can range between values of about 0.35 and 1.00. If the void ratio is more than unity the microstructure is collapsible or metastable. Generally speaking the larger the particles, the higher the strength, and deposits consisting of a mixture of different-sized particles are usually stronger than those that are uniformly graded (Table 1.6).

Densely packed sands are almost incompressible, whereas loosely packed deposits, located above the water table, are relatively compressible but otherwise stable. If the relative density of a sand varies erratically, this can give rise to differential settlement. Greater settlement is likely to be experienced in granular soils when foundation level is below the water table than when above. Settlement is relatively rapid in granular soils.

Table 1.6 Some typical properties of gravels, sands and silts

	Gravels	*Sands*	*Silts*
Relative density	2.5–2.8	2.6–2.7	2.64–2.66
Bulk density (t/m^3)	1.45–2.3	1.4–2.15	1.82–2.15
Dry density (t/m^3)	1.4–2.1	1.35–1.9	1.45–1.95
Porosity (%)	20–50	23–35	–
Void ratio	–	–	0.35–0.85
Liquid limit (%)	–	–	24–35
Plastic limit (%)	–	–	14–25
Coefficient of consolidation (m^2/yr)	–	–	12.2
Effective cohesion (kPa)	–	–	75
Shear strength (kPa)	200–600	100–300	–
Angle of friction (deg)	35–45	32–42	32–36

1.3.2 Silts and loess

The grains in a deposit of silt are often rounded with smooth outlines. This influences their degree of packing. The latter, however, is more dependent on the grain size distribution within a silt deposit, uniformly sorted deposits not being able to achieve such close packing as those in which there is a range of grain size. This, in turn, influences the values of void ratio (0.35–0.85) as well as the bulk and dry densities (Table 1.6).

Dilatancy is characteristic of fine sands and silts. The environment is all-important for the development of dilatancy since conditions must be such that expansion can take place.

Consolidation of silt is influenced by grain size, particularly the size of the clay fraction, porosity and natural moisture content. Primary consolidation may account for over 75% of total consolidation. In addition, construction settlement may continue for several months after completion because the rate at which water can drain from the voids under the influence of applied stress is slow.

Loess owes its engineering characteristics largely to the way in which it was deposited since this commonly has given it a metastable structure, in that initially the particles were loosely packed. The porosity of the structure is enhanced by the presence of fossil root-holes. The latter are lined with carbonate cement, which helps bind the grains together. This means that the initial, loosely packed structure is preserved and the carbonate cement provides some of the bonding strength of loess. However, the chief binder is usually the clay matrix. On wetting, the clay bond in many loess soils becomes soft, which can lead to the collapse of the metastable structure. The breakdown of the soil structure can occur under its own weight.

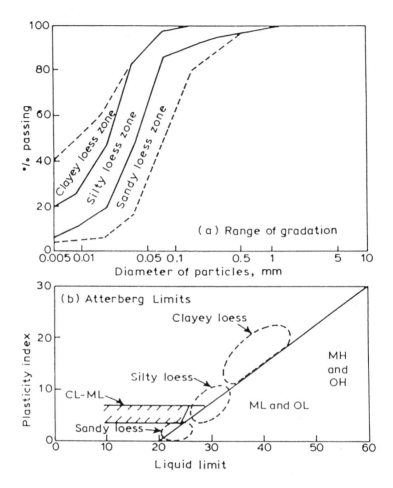

Figure 1.2 Particle size distribution and plasticity of Missouri river basin loess (after Clevenger, 1958, with permission of ASCE)

Loess deposits generally consist of 50–90% particles of silt size. In fact sandy, silty and clayey loess can be distinguished. The undisturbed densities of loess may range from around 1.2–1.36 t/m^3. If wetted (or reworked), the density of collapsible loess increases, sometimes to as high as 1.6 t/m^3. The liquid limit of loess averages about 30% (exceptionally liquid limits as high as 45% have been recorded), and their plasticity index ranges from about 4 to 9%, but averages 6% (Figure 1.2).

Normally loess possesses a high shearing resistance and can carry high loadings without significant settlement when natural moisture contents are low. For instance, natural moisture contents of undisturbed loess are generally around 10% and the supporting capacity of loess at this moisture content is high. However, the density of loess is the most important factor controlling its shear strength and settlement. On wetting, large settlements and low shearing resistance are encountered when the density of loess is below 1.30 t/m^3, whereas if the density exceeds 1.45 t/m^3 settlement is small and shearing resistance is fairly high.

Loess deposits are better drained (their permeability ranges from 10^{-5} to 10^{-7} m/s) than are true silts because of the fossil root-holes. Their permeability is appreciably higher in the vertical than in the horizontal direction.

1.3.3 Clay deposits

The principal minerals in a deposit of clay tend to influence its engineering behaviour. For example, the plasticity of a clay soil is influenced by the amount of its clay fraction and the type of clay minerals present since clay minerals greatly influence the amount of attracted water held in a soil. The undrained shear strength is related to the amount and type of clay minerals present in a clay deposit together with the presence of cementing agents. In particular, strength is reduced with increasing content of mixed-layer clay and montmorillonite in the clay fraction. The increasing presence of cementing agents, especially calcite, enhances the strength of the clay.

Geological age also has an influence on the engineering behaviour of a clay deposit. The porosity, water content and plasticity normally decrease in value with increasing depth, whereas the strength and elastic modulus increase.

The engineering performance of clay deposits is also affected by the total moisture content and by the energy with which this moisture is held. For instance, the moisture content influences their consistency and strength, and the energy with which moisture is held influences their volume change characteristics.

One of the most notable characteristics of clays from the engineering point of view is their susceptibility to slow volume changes that can occur independent of loading due to swelling or shrinkage. Differences in the period and magnitude of precipitation and evapotransportation are the major factors influencing the swell–shrink response of a clay beneath a structure. Generally kaolinite has the smallest swelling capacity of the clay minerals. Illite may swell by up to 15% but intermixed illite and montmorillonite may swell some 60–100%. Swelling in Ca montmorillonite is very much less than in the Na variety; it ranges from about 50 to 100%. Swelling in Na montmorillonite occasionally can amount to 2000% of the original volume. One of the

Introduction

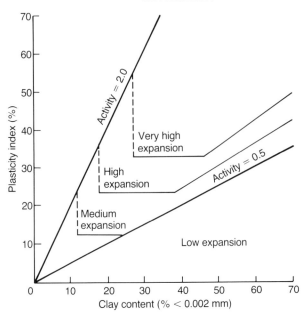

Figure 1.3 Estimation of the degree of expansiveness of a clay soil (after Williams and Donaldson, 1980)

most widely used soil properties to predict swell potential is the activity of a clay (Figure 1.3).

Volume changes in clays also occur as a result of loading and unloading which bring about consolidation and heave, respectively. When a load is applied to a clay soil its volume is reduced, this being due principally to a reduction in the void ratio. If such a soil is saturated, then the load is initially carried by the pore water which causes a pressure, the hydrostatic excess pressure, to develop. The excess pressure of the pore water is dissipated at a rate which depends upon the permeability of the soil mass, and the load is eventually transferred to the soil structure. Primary consolidation is brought about by a reduction in the void ratio. Further consolidation may occur due to a rearrangement of the soil particles. This secondary compression is usually much less significant. In clay soils, because of their low permeability, the rate of consolidation is slow. The compressibility of a clay is related to its geological history, that is, to whether it is normally consolidated or overconsolidated.

The heave potential arising from stress release depends upon the nature of the diagenetic bonds within the soil. For example, when an excavation is made in a clay with weak diagenetic bonds, elastic rebound causes immediate dissipation of some stored strain energy in the soil, this being manifested in a certain amount of heave.

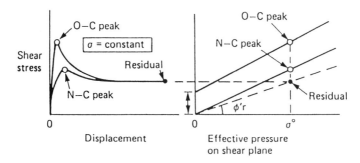

Figure 1.4 Peak strength and residual strength of normally (N–C) and over-consolidated (O–C) clays (after Skempton, 1964)

An overconsolidated clay is considerably stronger at a given confining pressure than a normally consolidated clay and tends to dilate during shear, whereas a normally consolidated clay consolidates. Hence, when an overconsolidated clay is sheared under undrained conditions negative pore water pressures are induced, the effective strength is increased, and the undrained strength is much higher than the drained strength – the exact opposite to a normally consolidated clay. When the negative pore-water pressure gradually dissipates the strength falls as much as 60 or 80% to the drained strength.

Skempton (1964) observed that when clay is strained it develops an increasing resistance (strength), but that under a given effective pressure the resistance offered is limited, the maximum value corresponding to the peak strength. If testing is continued beyond the peak strength, then, as displacement increases, the resistance decreases, again to a limiting value which is termed the *residual strength*. In moving from peak to residual strength, cohesion falls to almost, or actually, zero and the angle of shearing resistance is reduced to a few degrees (it may be as much as 10° in some clays). Under a given effective pressure, the residual strength of a clay is virtually the same whether it is normally consolidated or overconsolidated (Figure 1.4). Furthermore, the value of residual shear strength (ϕ'_r) decreases as the amount of clay fraction increases in a deposit. Not only is the proportion of detrital minerals important but so is that of the diagenetic minerals. The latter influence the degree of induration of a deposit of clay and the value of ϕ'_r can fall significantly as the ratio of clay minerals to detrital and diagenetic minerals increases.

The shear strength of an undisturbed clay is frequently found to be greater than that obtained when it is remoulded and tested under the same conditions and at the same water content. The ratio of the undisturbed to the remoulded strength at the same moisture content is termed the *sensitivity* of a clay. Clays with high sensitivity values have little or no strength after being disturbed. Sensitive clays generally possess high moisture contents, frequently with liquidity indices well in excess of unity. A sharp increase in moisture content may cause a great increase in sensitivity, sometimes with disastrous results. Heavily overconsolidated clays are insensitive.

Fissures in clays play an extremely important role in their failure mechanism and are characteristic of overconsolidated clays. For example, the strength along fissures in the clay is only slightly higher than the residual strength of the intact clay. Hence, the upper limit of the strength of fissured clay is represented by its intact strength while the lower limit corresponds to the strength along the fissures. The operational strength is, however, often significantly higher than the fissure strength. The ingress of water into fissures means that the pore-water pressure in the clay concerned increases, which in turn means that its strength is reduced.

Fissures in overconsolidated clays can have practical consequences in that the strength of the clays can be significantly reduced. For example, Skempton *et al.* (1969) summarized the shear strength parameters of London Clay (which is a heavily overconsolidated fissured clay) in terms of effective stress as follows:

1. Peak strength of intact clay: $c' = 31$ kPa, $\phi' = 20$;
2. 'Peak' strength on fissure and joint surfaces: $c' = 6.9$ kPa, $\phi' = 18.5$;
3. Residual strength of intact clay: $c'_r = 1.4$ kPa, $\phi'_r = 16$.

The greatest variation in the engineering properties of clays can be attributed to the degree of weathering they have undergone. Ultimately weathering, through the destruction of interparticle bonds, leads to a clay deposit reverting to a normally consolidated, sensibly remoulded condition. Higher moisture contents are found in the more weathered clay. This progressive degrading and softening is also accompanied by reductions in strength and deformation moduli with a general increase in plasticity.

1.3.4 Tropical soils

Ferruginous and aluminous clay soils are frequent products of weathering in tropical latitudes (Anon., 1990(a)). They are characterized by the presence of iron and aluminium oxides and hydroxides. Laterite is a residual ferruginous clay-like deposit which generally occurs below a hardened ferruginous crust or hardpan. During drier periods the water table is lowered. The small amount of iron that has been mobilized in the ferrous state by the groundwater is then oxidized, forming haematite or, if hydrated, goethite. The movement of the water table leads to the gradual accumulation of iron oxides at a given horizon in the

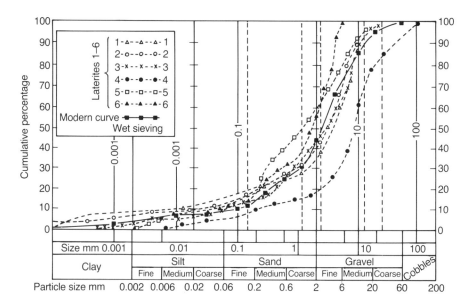

Figure 1.5 Grading curves of laterites (after Madu, 1977)

soil profile. A cemented layer of laterite is formed which may be a continuous or honeycombed mass, or nodules may be formed, as in laterite gravel.

Laterite hardens on exposure to air. Hardening may be due to a change in the hydration of iron and aluminium oxides. It commonly contains all size fractions from clay to gravel and sometimes even larger material (Figure 1.5). Values of common properties of laterite are given in Table 1.7. Such soils are of low to medium plasticity. The strength of laterite may decrease with increasing depth beneath the hardened crust

Table 1.7 Some common properties of laterites

Moisture content (%)	10–49
Liquid limit (%)	33–90
Plastic limit (%)	13–31
Clay fraction	15–45
Dry unit weight (kN/m^3)	15.2–17.3
Cohesion, c_u(kPa)	466–782
Angle of internal friction, ϕ_u (°)	28–35
Unconfined compressive strength (kPa)	220–825
Compression index	0.0186
Coefficient of consolidation (m^2/year)	262–599*
Young's modulus (kPa)	5.63×10^4

* For a pressure of 215 kPa

(where present). The latter has a low compressibility and settlement, therefore, is usually negligible.

Red earths or latosols are residual ferruginous soils in which oxidation readily occurs. Most of them appear to have been derived from the first cycle of weathering of the parent material. They differ from laterite in that they behave as a clay and do not possess strong concretions. They do, however, grade into laterite.

Black clays are typically developed on poorly drained soils in regions with well-defined wet and dry seasons, where the annual rainfall is not less than 1250 mm. Generally the clay fraction in these soils exceeds 50%, silty material varying between 20 and 40%, and sand forming the remainder. The organic content is usually less than 2%. The liquid limits of black clays may range between 50 and 100%, with plasticity indices of between 25 and 70%. The shrinkage limit is frequently around 10–12%. Montmorillonite is commonly present in the clay fraction and is the chief factor determining the behaviour of these clays. For instance, they undergo appreciable volume changes on wetting and drying due to the montmorillonite content. These volume changes, however, tend to be confined to an upper critical zone of the soil, which is frequently less than 2.5 m thick. Below this the moisture content remains more or less the same, for instance, around 25%.

Dispersive clay soils deflocculate in the presence of relatively pure water to form colloidal suspensions and therefore are highly susceptible to erosion and piping. Piping is initiated by dispersion of clay particles along desiccation cracks, fissures and root-holes. There is no threshold velocity for dispersive clay, the colloidal clay particles go into suspension even in quiet water. Hence retrogressive erosion can occur at very low pore-water flow velocities. Such soils contain a higher content of dissolved sodium in their pore water than ordinary soils (up to 12%, with pH values varying between 6 and 8). There are no significant differences in the clay contents of dispersive and non-dispersive soils, except that soils with less than 10% clay particles may not have enough colloids to support dispersive piping. Potentially dispersive soils frequently contain a moderate to high content of clay.

Dispersive erosion depends on the mineralogy and chemistry of the soil on the one hand, and the dissolved salts in the pore and eroding water on the other. The presence of exchangeable sodium is the main chemical factor contributing to dispersive clay behaviour. This is expressed in terms of the exchangeable sodium percentage (ESP = exchangeable sodium ÷ cation exchange capacity, %) where the units are given in meq./100 g of dry soil. Above threshold values of ESP of 10%, soils that have their free salts leached by seepage of relatively pure water are prone to dispersion. Soils with ESP values above 15% are highly dispersive. High ESP values and piping potential generally exist

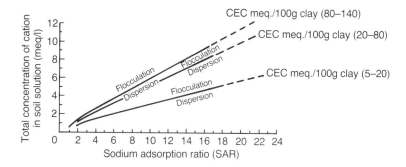

Figure 1.6 The influence of free salts on dispersion and flocculation as influenced by the colloidal composition of soils. SAR = Na/[(Ca + Mg)/2]

in soils in which the clay fraction is composed largely of smectitic and other 2:1 clays. Some illitic soils are highly dispersive. High values of ESP and high dispersibility are rare in clays composed largely of kaolinites.

The main property of clay soils governing their susceptibility to dispersion is the total content of dissolved salts in the water. The lower the content of dissolved salts in the water, the greater the susceptibility of sodium-saturated clay to dispersion. There is a threshold value for total cation concentration (TCC) in the pore water (for a given ESP) above which the soil remains flocculated. Figure 1.6 shows the zones in which a soil of a given ESP can exist in either a dispersed or flocculated state, depending on the concentration of the salts in the pore water. A number of special tests are used to identify dispersive soils. They have been reviewed by Sherard *et al.* (1976) and Gerber and Harmse (1987).

Calcareous silty clays are important types of soil in arid and semi-arid areas. These silty clays are light to dark brown in colour. They normally are formed by deposition in saline or lime-rich waters. These soils possess a stiff to hard desiccated clay crust, referred to as duricrust, which may be up to 2 m thick, and which overlies moist soft silty clay.

In arid and semi-arid regions the evaporation of moisture from the surface of the soil may lead to the precipitation of salts in the upper layers. The most commonly precipitated material is calcium carbonate. These caliche deposits are referred to as calcrete. The development of calcrete is inhibited beyond a certain aridity since the low precipitation is unable to dissolve and drain calcium carbonate towards the water table. Consequently in arid climates gypcrete may take the place of calcrete.

The hardened calcrete crust may contain nodules of limestone or be more or less completely cemented (this cement may, of course, have been subjected to differential leaching). As the carbonate content increases it first occurs as scattered concentrations of flaky habit, then as hard concretions. Once it exceeds 60%, the concentration becomes continuous. The calcium carbonate in calcrete profiles decreases from top to base, as generally does the hardness.

1.3.5 Tills and other glacial deposits

Till is usually regarded as being synonymous with boulder clay. It is deposited directly by ice while stratified drift is deposited in melt waters associated with glaciers. The character of till deposits varies appreciably and depends mainly on the lithology of the material from which it was derived. The underlying bedrock material usually contributes up to about 80% of basal or lodgement tills. Lodgement till is plastered on to the ground beneath a moving glacier in small increments as the basal ice melts. Because of the overlying weight of ice such deposits are overconsolidated.

The proportion of silt and clay size material is relatively high in lodgement till (e.g. the clay fraction varies from 15 to 40%). Lodgement till is commonly stiff, dense and relatively incompressible. Hence it is practically impermeable. Fissures are frequently present in lodgement till, especially if it is clay matrix dominated.

Ablation till accumulates on the surface of the ice when englacial debris melts out, and as the glacier decays the ablation till is slowly lowered to the ground. It is therefore normally consolidated. Because it has not been subjected to much abrasion, ablation till is characterized by abundant large stones that are angular, the proportion of sand and gravel is high and clay is present only in small amounts (usually less than 10%). Because the texture is loose, ablation till can have an extremely low *in-situ* density. Since ablation till consists of the load carried at the time of ablation it usually forms a thinner deposit than lodgement till.

Tills are frequently gap graded, the gap generally occurring in the sand fraction (Figure 1.7). Large, often very local, variations can occur in the grading of till. The range in the proportions of coarse and fine fractions in tills dictates the degree to which the properties of the fine fraction influence the properties of the composite soil. The variation in the engineering properties of the fine soil fraction is greater than that of the coarse fraction, and this often tends to dominate the engineering behaviour of the till.

The consistency limits of tills are dependent upon water content, grain size distribution and the properties of the fine-grained fraction. Generally, however, the plasticity index is small and the liquid limit of tills decreases with increasing grain size. Dense, heavily overconsolidated till is relatively incompressible and when loaded

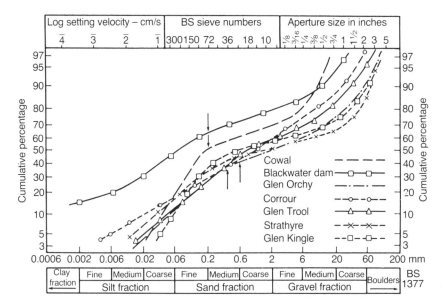

Figure 1.7 Typical gradings of some Scottish morainic soils (after McGown, 1971)

undergoes very little settlement, most of which is elastic. For the average structure such elastic compressions are too small to be of concern and therefore can be ignored.

Opening up and softening along fissures gives rise to a rapid reduction of undrained shear strength along the fissures. Deformation and permeability are also controlled by the nature of the fissures. Fissures tend to be variable in character, spacing, orientation and areal extent.

Deposits of stratified drift are often subdivided into two categories, namely, those which develop in contact with the ice – the ice contact deposits – and those which accumulate beyond the limits of the ice, forming in streams, lakes or seas – the proglacial deposits.

Outwash fans range in particle size from coarse sands to boulders. When they are first deposited their porosity may be anything from 25 to 50% and they tend to be very permeable. The finer silt–clay fraction is transported further downstream. Kames, kame terraces and eskers usually consist of sands and gravels.

The most familiar proglacial deposits are varved clays. The thickness of the individual varve is frequently less than 2 mm, although much thicker layers have been noted in a few deposits. Generally the coarser layer is of silt size and the finer of clay size. Varved clays tend to be normally consolidated or lightly overconsolidated.

The range of liquid limits for varved clays tends to vary between 30 and 80%, while that of plastic limit often varies between 15 and 30%. Hence they are inorganic silty clays of medium to high plasticity or compressibility. In some varved clays the natural moisture content is near the liquid limit so they are medium-sensitive clays. The effective stress parameters of apparent cohesion and angle of shearing resistance frequently range from 5 to 19.5 kPa and 22 to 25° respectively.

The material of which quick clays are composed is predominantly smaller than 0.002 mm but many deposits seem to be very poor in clay minerals, containing a high proportion of ground-down, fine quartz. Particles, whether aggregations or individual minerals, are rarely in direct contact, being linked generally by bridges of fine particles.

Quick clays generally exhibit little plasticity, their plasticity index generally varying between 8 and 12%. Their liquidity index normally exceeds 1, and their liquid limit is often less than 40%. The most extraordinary property possessed by quick clays is their very high sensitivity.

1.3.6 Organic soils: peat

Peat is an accumulation of partially decomposed and disintegrated plant remains which have been fossilized under conditions of incomplete aeration and high water content. Physico-chemical and biochemical processes cause this organic material to remain in a state of preservation over a long period of time.

Macroscopically, peaty material can be divided into three basic groups, namely, amorphous granular, coarse fibrous and fine fibrous peat. The amorphous granular peats have a high colloidal fraction, holding most of their water in an adsorbed rather than free state. In the other two types the peat is composed of fibres, these usually being woody. In the coarse variety a mesh of second-order size exists within the interstices of the first-order network, while in fine fibrous peat the interstices are very small and contain colloidal matter.

The mineral material in peat is usually quartz sand and silt. In many deposits the mineral content increases with depth. The amount of mineral content influences the engineering properties of peat.

The void ratio of peat ranges from about 9 for dense amorphous granular peat, up to 25 for fibrous types with high contents of sphagnum. It tends to decrease with depth within a peat deposit. Such high void ratios give rise to a phenomenally high water content and most of the peculiarities of peat are attributable to its moisture content. This varies according to the type of peat; it may be as low as 500% while values exceeding 3000% occur. The amount of shrinkage that can occur in peat generally ranges between 10 and 75% of the original volume, and it can involve reductions in void ratio from over 12 down to about 2.

Dry densities of drained peat fall within the range 65–120 kg/m^3. The dry density is influenced by the mineral content, and higher values than those mentioned can be obtained when peats possess high mineral residues.

Apart from its moisture content and dry density, the shear strength of a peat deposit appears to be influenced by its degree of humidification and its mineral content. As both these factors increase, so does the shear strength. Conversely, the higher the moisture content of peat, the lower is its shear strength.

In an undrained bog the unconfined compressive strength may be negligible, the peat possessing a consistency approximating to that of a liquid. The strength is increased by drainage to values between 20 and 30 kPa and the modulus of elasticity to between 100 and 140 kPa. When loaded, peat deposits undergo high deformations but their modulus of deformation tends to increase with increasing load. If peat is very fibrous it appears to suffer indefinite deformation without planes of failure developing, whereas failure planes nearly always form in dense amorphous peats.

Differential and excessive settlement is the principal problem confronting the engineer working on a peaty soil. When a load is applied to peat, settlement occurs because of the low lateral resistance offered by the adjacent unloaded peat. Serious shearing stresses are induced even by moderate loads. Worse still, should the loads exceed a given minimum, then settlement may be accompanied by creep, lateral spreading or, in extreme cases, by rotational slip and upheaval of adjacent ground. At any given time the total settlement in peat due to loading involves settlement with and without volume change. Settlement without volume change is the more serious for it can give rise to the types of failure mentioned. What is more, it does not enhance the strength of peat.

1.3.7 Fills

Because suitable sites are becoming scarce in urban areas, the construction of buildings on fill or made-up ground has assumed a greater importance. A wide variety of materials is used for fills, including domestic refuse, ashes, slag, clinker, building waste, chemical waste, quarry waste and all types of soils. The extent to which an existing fill will be suitable as a foundation depends largely on its composition and uniformity. In the past the control exercised in placing fill has frequently been insufficient to ensure an adequate and uniform support for structures immediately after placement. Consequently a time interval had to be allowed prior to building so that the material could consolidate under its own weight. Although this may be suitable for small, lightly loaded buildings it is unsatisfactory for more heavily loaded structures that can give rise to substantial settlement.

The time taken for a fill to reach a sufficient degree of natural consolidation so that it becomes suitable for a foundation depends on the nature and thickness of the fill, the method of placing and the nature of the underlying ground, especially the groundwater conditions. The best materials in this respect are obviously well graded, hard and granular (Figure 1.8). By contrast, fills containing a large proportion of fine material may take a long time to settle. Generally rock fills will settle 2.5% of their thickness, sandy fills about 5% and cohesive material around 10%. The rate of settlement decreases with time but in some cases it may take 10–20 years before movements are reduced within tolerable limits for building foundations. In coarse-grained soils the larger part of movement generally occurs within the first two years after the construction of the fill, and after five years settlements are usually

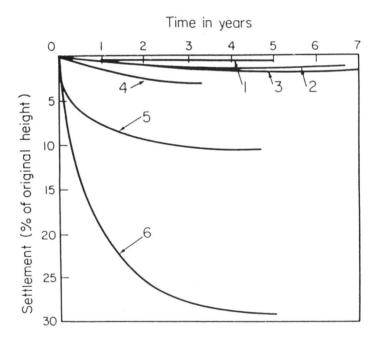

Figure 1.8 Observations of the settlement of various types of fill due to consolidation under its own weight (after Meyerhof, 1951). Description of curves: 1, well-graded sand, well compacted; 2, rockfill, medium state of compaction; 3, clay and chalk, lightly compacted; 4, sand, uncompacted; 5, clay, uncompacted; 6, mixed refuse, well compacted

Table 1.8 Municipal waste materials incorporated in fills (after Sowers, 1973)

Material	Characteristics as fill
1. Garbage: food, waste	Wet. Ferments and decays readily. Compressible, weak
2. Paper, cloth	Dry to damp. Decays and burns. Compressible
3. Garden refuse	Damp. Ferments, decays, burns. Compressible
4. Plastic	Dry. Decay resistant, may burn. Compressible
5. Hollow metal, e.g. drums	Dry. Corrodable and crushable
6. Massive metal	Dry. Slightly corrodable. Rigid
7. Rubber, e.g. tyres	Dry. Resilient, burns, decay resistant. Compressible
8. Glass	Dry. Decay resistant. Crushable and compressible
9. Demolition timber	Dry. Decays and burns. Crushable
10. Building rubble	Damp. Decay resistant. Crushable and erodable
11. Ashes, clinker and chemical wastes	Damp. Compressible, active chemically and partially soluble

very small. The minimum time that should elapse before development takes place on an opencast backfill should be 12 years after restoration is complete. Frequently, poorly compacted old fills continue to settle for years due to secondary consolidation.

Waste disposal or sanitary land fills are usually very mixed in composition (Table 1.8) and suffer from continuing organic decomposition and physico-chemical breakdown which may leave voids. Methane and hydrogen sulphide are often produced in the process, and accumulations of these gases in pockets in fills have led to explosions. The production of leachate is another problem. Some material such as ashes and industrial wastes may contain sulphates and other products which are potentially damaging to concrete. The density of waste disposal fills varies from about 120 to 300 kg/m^3 when tipped. After compaction the density may exceed 600 kg/m^3. Moisture contents range from 10 to 50% and the average specific gravity of the solids from 1.7 to 2.5. Settlements are likely to be large and irregular. The initial settlement of waste disposal fills is rapid and is due to a reduction in the void ratio. It takes place with no build up of pore-water pressure. Settlement continues due to a combination of secondary compression (material disturbance) and physico-chemical and biochemical action.

1.3.8 Coarse colliery discard

There are two types of colliery discard, namely, coarse and fine. Coarse discard consists of run-of-mine material and reflects the various rock types that are extracted during mining operations. It contains varying amounts of coal that has not been separated by the preparation process. Fine discard consists of either slurry or tailings from the washery, which is pumped into lagoons. Some tips, particularly those with relatively high coal contents, may be partly burned or burning and this affects their composition and, therefore, their engineering behaviour.

The majority of tip material is essentially granular. Often most of it falls within the sand range, but significant proportions of gravel and cobble range may also be present. Owing to breakdown, older and surface materials tend to contain a higher proportion of fines than that occurring within a tip. The moisture content of coarse discard increases with increasing content of fines, and generally falls within the range 5–15%. Tip material shows a wide variation in bulk density and may, in fact, vary within a spoil heap. Low densities are mainly a function of low specific gravity.

As far as effective shear strength of coarse discard is concerned, ϕ' usually varies from 25 to 45°. The angle of shearing resistance, and therefore the strength, increases in spoil that has been burned. With increasing content of fine coal, on the other hand, the angle of shearing resistance is reduced. The shear strength of colliery spoil, and therefore its stability, is dependent upon the pore pressures developed within it. These are likely to be developed where there is a high proportion of fine material, which reduces the permeability below 5×10^{-7} m/s.

Oxidation of pyrite within tip waste is governed by access of air. However, the highly acidic oxidation products that result may be neutralized by alkaline materials in the waste. The sulphate content of weathered, unburned colliery waste is usually high enough to warrant special precautions in the design of concrete structures which may be in contact with the discard.

Spontaneous combustion of carbonaceous material, frequently aggravated by the oxidation of pyrite, is the most common cause of burning spoil. The problem of combustion has sometimes to be faced when reclaiming old tips. Spontaneous combustion may give rise to subsurface cavities in spoil heaps and burned ashes may also cover zones which are red hot to appreciable depths. When steam comes in contact with red-hot carbonaceous material, watergas is formed, and when the latter is mixed with air it becomes potentially explosive. Explosions may occur when burning spoil heaps are being reworked and a cloud of coal dust is formed near the heat surface.

Noxious gases are emitted from burning spoil. These include carbon monoxide, carbon dioxide, sulphur dioxide and, less frequently, hydrogen sulphide. Carbon monoxide may be present in potentially lethal concentrations. Sulphur gases are usually not present in high concentrations.

2

Exclusion techniques

Groundwater frequently provides one of the most difficult problems during excavation. Not only does water make working conditions difficult – for example, vehicles may become bogged down, muck becomes more difficult to handle – but flow of water into an excavation can lead to erosion and failure of the sides. Collapsed material has to be removed and the damage made good. Instability of the floor may arise when deep excavations are made into sandy ground. This is caused by upward seepage pressure of water, and piping and boiling results when the upward seepage velocity is high enough to suspend the individual grains of sand. Piping and boiling or quick conditions can also occur in silt. On the other hand, the velocity of groundwater flow in clay soils is usually so slow as not to present erosion problems. Some of the worse conditions are met in excavations that have to be taken beneath the water table.

Artesian conditions, in particular, can cause serious trouble in excavations and, therefore, if such conditions are expected it is essential that both the position of the water table and piezometric pressures should be determined before work commences. Otherwise excavations which extend close to strata under artesian pressure may be severely damaged due to blow-outs taking place in their floors. Such action may also cause slopes to fail and, in fact, has led to the abandonment of sites. Sites at which such problems are likely to be encountered should be dealt with prior to and during excavation, by employing exclusion or dewatering techniques.

In order to help select the type of cut-off required for groundwater control about an excavation, a number of basic questions have to be answered. First, is a cut-off the best method of groundwater control at the site concerned? Secondly, if it is, then which cut-off technique is best suited to the site conditions? Thirdly, can the cut-off be temporary – that is, remain in existence for the construction period only – or has it to be permanent? Fourthly, can the cut-off serve any other function?

Methods of forming such barriers include sheet piling, contiguous bored and secant pile walls, bentonite cut-off walls, geomembrane barriers, concrete diaphragm walls, grout curtains and panels, and ice walls (Bell and Mitchell, 1986). The economy of providing a barrier to exclude groundwater depends on the existence of an impermeable stratum beneath the excavation to form an effective natural cut-off for the barrier. If such a stratum does not exist or if it lies at too great a depth to be used as a cut-off, then upward seepage may occur which, in turn, may give rise to instability at the base of the excavation (Figure 2.1). In these circumstances the barrier will not be totally effective unless it can be extended horizontally beneath the excavation. The only methods of forming a horizontal barrier are by grouting or freezing.

In urban areas the presence of existing buildings and the high price of land frequently dictate that construction be carried out in the least possible space. Where surrounding property has to be safeguarded it will usually be more appropriate to provide a barrier or cut-off around an excavation so as to prevent the inflow of water while maintaining the water table outside at its normal level, rather than adopt a dewatering technique. By keeping the water table at its normal level, settlement problems are not as likely to arise as they would if a dewatering technique was used. Exclusion techniques can be used with dewatering techniques.

Perimeter exclusion walls should be capable of construction through soft, wet ground; these walls should be economical in installation cost and should be capable of being installed without any adverse effects on the stability of adjacent buildings. If possible, they should eliminate the need for temporary work during construction of the permanent works. Any movement of the barrier can adversely affect both the structure

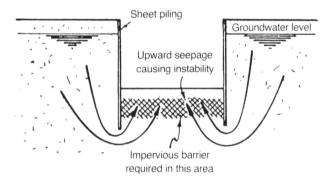

Figure 2.1 Sheet piling and the possibility of instability at the base of an excavation due to upward seepage pressure.

to be incorporated within the excavation and the stability of existing buildings immediately adjacent to it. Accordingly, measurements should be taken of the constructed walls and adjacent properties to check for any movements.

2.1 SHEET PILES

Sheet piling may be composed of steel, timber or concrete piles, each pile being linked to the next to form a continuous wall. The integrity of sheet-pile walls depends on the interlock between individual piles. Steel sheet piles are most widely used (Figure 2.2) since they represent the simplest and cheapest method of forming an impermeable barrier, especially if the piles can be extracted for re-use. By contrast, precast reinforced concrete piles nowadays are seldom used.

Steel sheet piles are used in temporary structures such as strutted excavations and cofferdams, as well as for permanent use, notably in retaining walls and bulkheads. With all sections of pile or types of steel, higher stresses are permissible when used in temporary works rather than for permanent use. Sheet piling, when installed before excavation takes place, provides an effective cut-off, as well as affording ground support in that the piles are used to resist horizontal pressures developed by soil and pore water. The piling derives its stability from the horizontal support provided by the ground into which it is driven, and from any anchors, ties or struts located near the top of the piles.

Steel sheet piling is most effective as a cut-off when driven into an underlying impermeable bed of clay. Should a thick deposit of sand exist below the clay, the differential head should be monitored by a piezometer (installed outside the piling) to ensure that the thickness of the bed of clay can offer adequate resistance. If it cannot, then pressure relief wells must be provided.

Steel sheet piling generally is not used in ground containing numerous boulders such as some tills, because the piles become difficult or impossible to drive and there is a risk that they might tear. One technique which has proved to be effective in such conditions involves excavating the boulders in a slurry trench, backfilling with sand, and then driving the piles. If the sand is thoroughly mixed before placement, the effectiveness of the piling as a cut-off is enhanced.

The pile section to be selected depends on height of ground to be retained, soil and groundwater conditions, and on whether the piling is tied back or strutted. The shape of the pile cross-section also is designed to make the wall capable of resisting bending. The section has to be capable of being driven into the soil to the desired depth and, in this context, the shape of the cross-section provides stiffness.

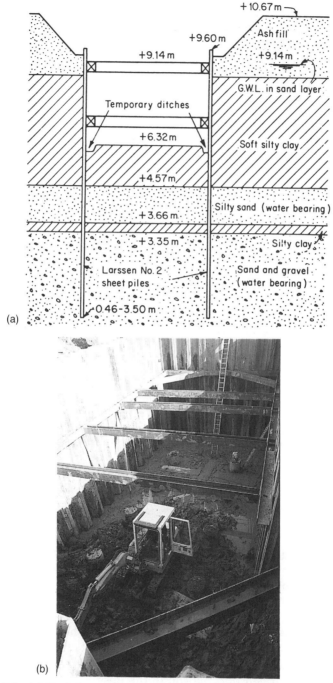

+10.67 m

Ash fill

+9.60 m

+9.14 m

+9.14 m

+9.14 m

G.W.L. in sand layer

Temporary ditches

+6.32 m

Soft silty clay

+4.57 m

+3.66 m

Silty sand (water bearing)

+3.35 m

Silty clay

Larssen No. 2
sheet piles

Sand and gravel
(water bearing)

0.46–3.50 m

(a)

(b)

Figure 2.2 (a) Steel sheet piling with bracing in variable soil conditions; (b) steel sheet piles used for forming a caisson in alluvial soils, near Kings Lynn, Norfolk

The most critical parts of a sheet-pile wall for groundwater exclusion are the interlocks which provide the connection between piles, thereby forming a continuous wall. Such a wall generally is sufficiently watertight for most practical purposes. However, if piles come out of interlock during driving this means that the effectiveness of the cut-off is impaired. Moreover, when the sheet piling remains interlocked, the cut-off is still of limited effectiveness until the sheet is stressed, wedging the adjacent piles into tight contact. In fact leakage can be quite high when a row of steel sheet piles, acting as a cut-off, is unstressed. It is therefore usually necessary to seal the interlocks by using heavy grease before installation, or by grouting.

Since a much larger amount of penetration below excavation level is required for a cantilever wall than an anchored wall, it is imperative to ensure that the section of piling finally calculated on bending moment considerations is large enough to withstand the vertical stresses developed during driving. In the former type of wall the depth of penetration should be at least equal to the height of the sheet piling above excavation level.

When steel sheet piling extends into undisturbed soil, even though it goes below the water table, corrosion tends to be very slight. Hence it is generally unnecessary to apply a protective coating to the piles. On the other hand, some corrosion may occur above the water table in disturbed soil and it is then necessary to apply a protective coating. Normally two coatings of a bituminous paint are given to steel sheet piles before installation.

Sheet piling about an excavation frequently requires support. In such cases sheet piling can be driven in advance of excavation. As excavation proceeds, waling is placed against the sheet piling and struts are placed across the excavation and wedged against the waling. Alternatively, soldier beams are driven at intervals along the line of excavation, and as the excavation proceeds wooden sheeting planks are inserted horizontally against the ground and are supported by the soldier beams. Tiebacks or anchors also may be used as support.

Steel sheet piling is used most frequently for anchored retaining walls. These normally range in height between 4.5 and 12 m, depending on the soil conditions. Because of its structural strength, watertightness and ability to be driven to appreciable depth in most soils, steel sheet piling is also widely used for cofferdams. However, in some soils – notably fine sands and silts beneath the water table – there is a danger of quick conditions developing if the critical gradient is exceeded as ground is removed. Piping sometimes occurs along steel sheet piling, especially when it is driven with the aid of jetting. In such situations consideration should be given to lowering the water table below the base of the excavation.

2.2 CONTIGUOUS BORED PILES AND SECANT PILES

Contiguous bored piles are frequently associated with both shallow and deep excavations for basements to buildings and with cofferdam work. This is particularly the case in urban areas where noise or the effect of installation on adjacent property are important, as well as in industrial complexes where access, headroom or restrictions on vibration may make other exclusion methods such as steel sheet piling less acceptable. Contiguous piles can be designed to cantilever vertically, to be strutted or to be anchored back. It is usual to employ ground anchors, either temporarily or as part of the final design, to eliminate the need for struts which limit the working space within an excavation and also to reduce the required pile diameter and amount of reinforcement within the pile. Bored piles possess a number of advantages. For example, their length can be varied to suit the ground conditions. Normally temporary casings have to be used during construction of contiguous bored piles although in clay soils only a short length of casing generally is necessary to provide a seal at the top of the clay. Contiguous bored piles normally are best suited to cohesive soils. By contrast they may be impossible to install in saturated loosely packed silts or sands owing to the soil undergoing liquefaction. Moreover a loss of ground may occur when sinking piles in groups in non-cohesive soils, with possible settlement of adjacent structures. In soils where the boring is slow or difficult, owing, for example, to the presence of boulders, the amount of overbreak resulting from driving and boring out casings at very close centres in line, is likely to be greater than that arising from single piles at wide spacing. Difficulties can arise when constructing bored piles in water-bearing or in squeezing ground. For instance, if the bottom of the casing is lifted above the base of the concrete pile while concreting operations are still proceeding, with the casing being withdrawn in a number of lifts, then water may enter the borehole and thereby lower the strength of the concrete. Under artesian conditions water may pipe up the pile shaft and in so doing remove cement. Squeezing ground may give rise to 'necking' in a pile.

When contiguous bored piles are used for retaining walls, the piles are installed either in a single or double row and are positioned such that they are touching or are in very close proximity to each other. The piles are reinforced to resist the applied bending moments and shear forces. There is much flexibility as far as the choice of pile diameter is concerned. However, the circular section of the pile generally means that more main reinforcement per metre run of wall is required than in the constant section of a diaphragm wall in order to produce an equivalent moment of resistance (North-Lewis and Lyons, 1975).

Presetting a number of casings in advance of the piling ensures the correct spacing of the piles in the line. The depth to which the casing extends depends on the ground conditions. For instance, where the piles are in clay, only a short length of casing generally is necessary to provide a seal in the top of the clay. Boring can then continue in an uncased hole. The use of temporary casing leads to gaps between piles. Such gaps normally should not exceed 75 mm at the head of the pile, and the verticality of the pile shafts should remain within the commonly accepted tolerances for bored piles (for example, 1 in 75).

The concrete of contiguous bored piles is placed directly into the pile shaft through a short trunking and hopper for dry conditions, and by the tremie technique under water. Alternate piles are constructed and the concrete allowed to harden before boring the intermediate piles. The holes are excavated by a grabbing rig or power auger.

Where water cannot be removed from a borehole either by baling or pumping, then the hole has to be cased from top to bottom to prevent collapse of the sides and a concrete plug is formed, beneath the water, at the base of the hole. Water is pumped from within the casing when the concrete plug has hardened sufficiently. Prior to filling the casing with concrete, it should be gently turned and lifted slightly to release it from the plug (Tomlinson, 1986).

When the concrete has hardened, the casing is then removed. Alternatively, concreting under water from a tremie pipe can take place over the complete length of the borehole. Such construction may not be able to avoid the formation of weak zones in the pile.

Where contiguous bored piles pass through water-bearing strata various grouting techniques can be employed to seal the joints. Grouting of joints is carried out before the excavation begins. The watertightness of the structure can be further improved by grouting the joints on the exposed face after excavation.

Watertightness also can be achieved by using interlocking secant piles. Alternate piles are first installed, but at a closer spacing than in the contiguous method (Figure 2.3(a)). Then the intermediate piles are installed by boring out the soil between alternate pairs, followed by chiselling a groove down the sides of the pile shafts. Lastly, concrete is placed in the holes, including the grooves, thereby forming a fully interlocking watertight wall.

A more recent method of installing secant pile walls is to form alternative bentonite–cement piles and reinforced concrete piles which interlock to provide a continuous wall. The bentonite–cement piles are constructed first, then the reinforced concrete piles are constructed midway between them. Parts of the bentonite–cement piles are bored out in order to form the reinforced concrete piles and in this way the interlock is achieved. The bentonite–cement piles are constructed in two

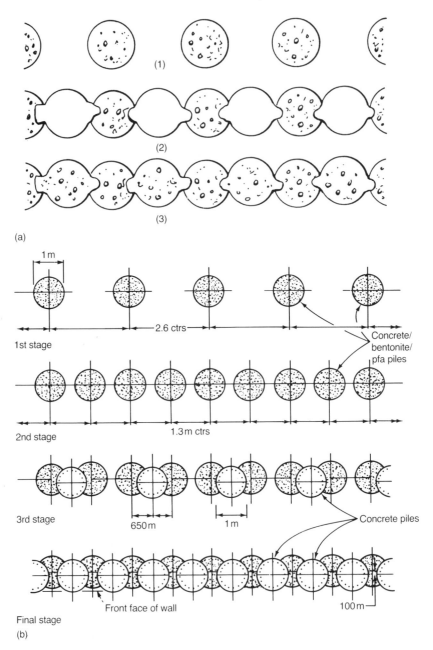

Figure 2.3 (a) Sequence of installation of secant piles: (1) initial piles emplaced; (2) groove chiselled in initial piles; (3) alternate piles cast. (b) Four stages in the construction of a secant pile wall using bentonite-cement and reinforced concrete piles (after Anon., 1985 (a)).

sequences. First, piles are formed at every second position (i.e. at double their spacings). Then after the first sites have set a second row of piles is constructed midway between the first. The reinforced concrete piles are emplaced when the bentonite–cement sites have gained sufficient strength for boring to proceed (Figure 2.3(b)). This is usually at least 36 hours after the bentonite–cement piles have been constructed.

2.3 SLURRY TRENCHES

Slurry trenches are used extensively as a means of groundwater cut-off. For example, they have been used to control seepage beneath dams and to contain groundwater pollution, notably from sanitary landfills and hazardous waste impoundments. Slurry trenches can achieve permeabilities of the order of 10^{-9} m/s.

There are two types of slurry trenches, namely, soil–bentonite and cement–bentonite. In the soil–bentonite cut-off process, the bentonite slurry is displaced by a soil–bentonite mixture similar in consistency to high-slump concrete. The soil–bentonite backfill consists of excavated soil, with the addition of material from a borrow pit mixed with bentonite mud, and forms a low permeability, highly plastic cut-off wall. In the cement–bentonite cut-off process cement is added to a fully hydrated bentonite slurry. The addition of the cement causes the slurry to harden, giving a strength comparable to that of stiff clay. At suitable sites, where slurry trenches have been used for basement, precast concrete panels have been lowered into slurry trenches and exposed by subsequent excavation to form the permanent basement walls (Figure 2.4). Self-setting slurry is used.

Impermeability is the most important property of a cut-off wall. There are three factors which account for the impermeability of a slurry trench, namely, the 'grouted' zone, the filter cake and the backfill. Bentonite initially may permeate the pores of the adjacent soil, depending upon the size distribution of the pores. As bentonite is thixotropic, when left undisturbed it gels and thereby seals the pores. The distance to which the soil is affected can range from almost zero in dense clays to a metre or so in loosely packed sands and gravels. A filter cake of loosely packed bentonite particles forms after a relatively short period on the sides of the trench during excavation; this results from the filtrate as the fluid escapes into the surrounding soil. This filter is usually a few millimetres in thickness and has a permeability as low as 10^{-11} m/s. Hence, it acts as a watertight membrane. The nature of the backfill has been referred to above. Permeability of the backfill can be as low as 10^{-9} m/s.

The chief factors affecting the permeability of a cement–bentonite slurry are the cement/water ratio, the bentonite/water ratio and the mechanical procedure in making panel connections between the fresh

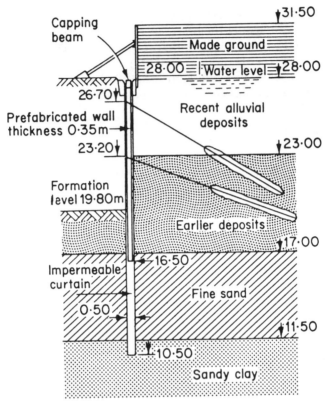

Figure 2.4 Precast concrete panels placed in slurry trench to form part of basement wall which is anchored back

cement–bentonite slurry and the set cement–bentonite. It is important that the bentonite be fully hydrated before any cement is added. This means that the cement/water ratio is the factor which controls the ultimate permeability of the backfill (generally between 1:4 and 1:5). The permeability of a cement–bentonite slurry cut-off wall can be as low as 10^{-9} m/s. Activated sodium bentonites have large swelling potential and are stable with cement. Between 35 and 45 parts bentonite mixed with between 160 and 220 parts cement, and water, according to Schweitzer (1989) gives a 'plastic' wall with an ultimate unconfined compressive strength of 1 MN/m^2 or over. This should allow a slurry wall to deform under loading without cracks developing.

Because it must have a low permeability, the continuity of slurry trench excavation is important. The depth of a slurry trench is controlled in many cases by the ground conditions, notably the depth

of an impermeable formation, and the type of barrier to be constructed. The minimum depth of penetration into an impermeable
formation at the base of the cut-off wall depends upon the nature of
the formation. For instance, if this is competent impervious rock, then
only a very small penetration may prove satisfactory. On the other
hand, an excavation into clay may need to go to a depth of 1 m or so.
The width is governed by the required permeability of the cut-off wall,
the head of water across the wall, the size of the excavation equipment
available, and the materials that form the wall. However, because the
filter cake provides the principal barrier to groundwater movement,
the width of the trench is not a major factor. In the case of soil–
bentonite backfill slurry walls, Millet and Perez (1981) suggested that
the wall should have thicknesses varying from 1.5 to 2.3 m. The
increased shear strength of the backfill forming cement–bentonite
slurry trench cut-off walls allows them to be thinner, their thickness
ranging from 0.6 to 0.9 m. Such widths are satisfactory for depths of
up to 30 m of hydrostatic head. Beyond this the ability of the cement–
bentonite cut-off wall to withstand hydrofracturing should be determined. The deviation from verticality is only important insofar as it
affects the continuity and integrity of the wall.

A major factor controlling the deformability of cut-off walls is the
properties of the materials of which they are composed. Soil–
bentonite walls are quite deformable and do not have problems with
regard to cracking. The cement/water ratio has a dramatic effect on
deformability of cement–bentonite backfill. For instance, the higher
the cement/water ratio, the higher the strength, the more rigid and
hence less deformable is the eventual cement–bentonite wall. On the
other hand, the higher the bentonite/water ratio, the more flexible,
and the more deformable the wall may be.

Excavation is made under slurry using backhoes, clamshells,
draglines and special devices. The head of slurry should be kept to
1–1.5 m above the maximum anticipated groundwater level in the
trench. The method of excavation chosen is based primarily on the
width and depth of the trench, and the type of soil. More than one
method of excavation can be used, for instance, a dragline can be
employed to dig the upper part of a trench (down to around 21 m),
greater depths being removed by clamshell. Speed of construction is
increased if several alternate panels can be formed during the initial
hardening period, the grab returning to dig out the intermediate
panels which, being of lesser width, are cut into the primary panels.
There are no joints between panels, therefore stop-end tubes and
guide trenches are not required.

As excavation proceeds, sands and silts become suspended in the
slurry. When a sample of slurry obtained from the trench exceeds a

specific gravity of about 1.6, the slurry should be circulated through settling ponds or separators. This improves the quality of the filter cake on the sides of the trench and enables the backfill to settle into position satisfactorily.

Specifications for soil–bentonite backfill have involved a wide range of particle sizes from coarse to fine, with the bentonite content in the range of 2–4% by weight, and of the order of 10–20% fines. Silty sand provides the ideal material for this purpose. Clays are suitable, except that hard clays tend to remain as blocks and therefore may give rise to voids in the fill. Cobbles and boulders, as well as roots or other organic material, should be removed before placement of fill.

The backfill is mixed with slurry at the surface, using bulldozers or front-end loaders, before being placed in the trench. The initial backfilling operation for a soil–bentonite cut-off wall involves the backfill being placed at the bottom of the trench by a clamshell bucket. This continues until the backfill material reaches the top of the trench. Then the remainder of the backfill at the surface is moved by a bulldozer onto the backfill exposed in the trench. This action forces the material to slide down into the trench under its own weight, and continues until the entire excavation has been backfilled. There is always some consolidation and settlement of a soil–bentonite cut-off wall. The effects of consolidation on the backfill material at the top of the soil–bentonite cut-off wall may lead to the development of seepage paths.

When slurry trenches are used to control groundwater pollution from hazardous waste sites, it may be that some of the chemicals present are incompatible with bentonite. In such situations high-density polyethylene sheeting is placed in the trench prior to backfilling. Sheeting also can be used to form an impermeable membrane when excavated material is not readily suitable for use as backfill yet cannot be economically disposed of. The permeability of high-density polyethylene sheeting can be as low as 10^{-14} m/s. When a slurry trench cut-off is required with a permeability of less than 10^{-9} m/s, then such sheeting is used to enclose the backfill.

Ressi and Cavalli (1984) described the construction of a geo-membrane sheeting-enclosed cut-off wall (the Environwall). First, a narrow trench is excavated and filled with bentonite slurry to prevent the trench collapsing. The geomembrane sheeting is fabricated to form a U-shaped envelope fitting the dimensions of the trench and then ballast is placed within it so that it can sink into the slurry trench. After initial submergence into the trench, the envelope is filled with wet sand. A system of wells and piezometers can be installed in the sand to monitor water quality and piezometric pressure, respectively. If the sheeting is damaged, then the system will detect the leakage.

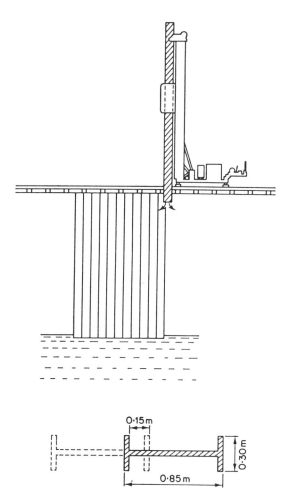

Figure 2.5 Beam method of forming a cut-off by driving or vibrating an I-beam. The length of the I-beam has to be at least equal to the depth of the cut-off being constructed

Wellpoints can be used to abstract this water and to maintain the net gradient into the liner.

In the beam method a cut-off wall is constructed by forming a narrow cavity in the soil by repeatedly driving or vibrating an I-beam section within it. As the I-beam is extracted an impervious material, usually a mixture of bentonite and cement, is pumped into the cavity (Figure 2.5). Successive penetrations of the I-beam are overlapped to develop a continuous membrane (Boyes, 1975). In the driven pile system, a clutch of H-piles is forced into the soil along the line of advance, the end pile being withdrawn and redriven at the head end. Grouting, by a tube fixed to the web, is carried out during withdrawal. Walls are limited to a maximum of 20 m depth with these systems due to the danger of deviating during driving.

2.4 DIAPHRAGM WALLS

Diaphragm walls represent a method of constructing walls in the ground in a narrow trench filled with bentonite slurry. Diaphragm walls compare favourably in terms of watertightness, stiffness and mechanical strength, with cut-offs formed of steel sheet piling, precast piles or cast *in-situ* piles. They may be rigid (concrete) or plastic (concrete and bentonite mixture) when load-carrying capacity is not required. Alternatively, diaphragm walls may be used as load-bearing and retaining walls and in such cases they are reinforced by incorporating a steel cage. Diaphragm walls are usually more economical when used as part of the permanent load-bearing structure.

A diaphragm wall allows a small area to be isolated for excavation using the wall as a sort of cofferdam. In such instances the panels are constructed to form a polygon. Diaphragm walls are composed of panels that are formed either successively or alternatively. In the first case, panels are formed next to each other in line. In the second case, primary panels are formed at regular intervals and the wall is then completed by constructing secondary panels between those already in place. Obviously watertight joints are required. In ground where artesian pressure exists, even a slight tilt of the panels relative to each other or inadequate concreting could cause disastrous flooding in deep excavations. This is even more important if the situation is aggravated by tidal fluctuations.

Diaphragm walls possess several features suitable for use in built-up areas (Boyes, 1975). For example, they can be constructed immediately adjacent to columns subject to live loadings, to multi-storey buildings and to old foundations. Structural waterproof walls may be formed in advance of the main excavation, which permits above- and below-ground construction at the same time if necessary. In many applications the interior portions of concrete diaphragm walls can be left exposed, as in subway excavations or basement car parks. Furthermore, the technique is virtually vibrationless and quiet, and so offers minimum disturbance.

The subsurface conditions should be thoroughly investigated so that any potential problems during excavation attributable to the presence of pervious layers are revealed. In particular, the groundwater conditions around the trench should be understood. This includes seasonal variations, as well as potential for dramatic changes in groundwater levels caused by anomalous weather conditions. Artesian pressures, springs and artificial sources of water (such as potential broken sewers) should be recorded to avoid instability of the trench during excavation.

In the construction of a diaphragm wall, a trench is first excavated to a depth of a metre or so. The guide walls are then constructed. The guide walls enable excavation of the main trench to begin below levels where made ground, drains or footings and other surface obstructions might be met. They support the trench against heavy construction surcharge pressure at levels over which the slurry rises and falls, and protect the sides against wave action set up during digging or during the introduction of fresh slurry.

A panel of predetermined dimensions is excavated between the guide walls (Figure 2.6). The width of a panel usually varies between 0.5 and 1 m, although greater widths are excavated when required. Lengths of up to 5 or 6 m represent the maximum that can be concreted conveniently with one tremie pipe (Millet and Perez, 1981). Problems of supply and coordination sometimes arise when more than one tremie pipe is used.

Slurry is introduced into the panel and maintained at as high a level as possible without spilling onto adjacent ground, thereby offering maximum support to the sides (Anon., 1985(b)). Individual panels are excavated by first digging to the full depth at each end, then the central section is excavated. Each grab or drill excavates a full panel width. The depths required in most structural work present little difficulty. The unavoidable fluctuation of the level of the bentonite in the trench during digging must be taken into account. It is generally recommended that the level of the slurry in the trench must be at least 1.25 m above that of the surrounding water table. If this cannot be realized, then the trench may be dug within a raised embankment, the water table may be lowered by installing wellpoints or the slurry may be weighted by adding sand or powdered barite.

After the panel has been excavated, any slurry–soil sediment that has accumulated at the base of the trench is removed and replaced with fresh slurry or treated with chemicals. Frequently, cleaning the base of the trench by removing the sludge with a grab proves adequate.

In a reinforced diaphragm wall the reinforcement cage is lowered into place through the slurry and supported from the concrete guide walls of the trench. Reinforcement is made up into cages prior to placement.

The concrete is placed using a tremie pipe. Slurry is pumped from the trench as the concrete is placed, being removed at a similar rate. Pours should be completed in the minimum time (up to 30 m^3/h) and with minimum discontinuity in order to avoid embedment of stop-end tubes, blockages in the tremie pipe and flotations of the cage due to upward drag from stiff concrete. All the concrete for a panel must be poured

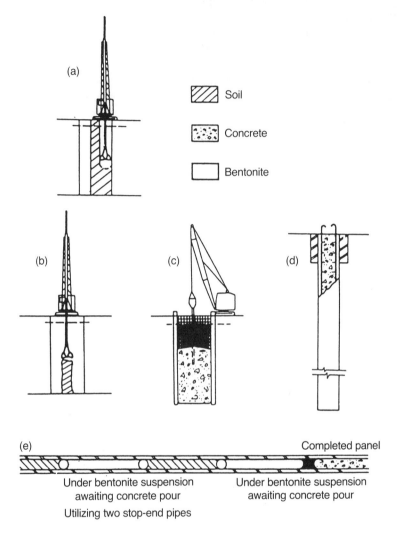

(a)

Soil

Concrete

Bentonite

(b)　　　(c)　　　(d)

(e)　　　　　　　　　　　　　　　　　Completed panel

Under bentonite suspension
awaiting concrete pour

Under bentonite suspension
awaiting concrete pour

Utilizing two stop-end pipes

Figure 2.6 Construction of a reinforced diaphragm wall. General procedure for excavation of panels during which time the excavation is kept filled with bentonite suspension; (a) first one end, then opposite end of panel is excavated to full depth; (b) third and last stage is the excavation of the centre panel. General procedure for concreting panels when the bentonite is displaced by concrete: (c) steel stop-end pipes and reinforcement cage positioned and concrete placed through tremie pipe; (d) section through completed panel showing guide walls and steel; (e) illustrating variety of uses of steel stop-end pipes as required

before setting or significant stiffening occurs. In practice this generally means 3–3.5 h.

As the concrete for each panel is tremied in, it moulds against the concave end of the previously cast panel at one end and against a circular steel stop-end placed temporarily at the other end. The knuckle-joint between adjacent panels provides sufficient connection for normal diaphragm wall work. Nevertheless, this simple butt joint between panels cannot prevent water penetration; however, significant leakages are rare. This is probably due to the presence of a thin layer of contaminated bentonite at the edges of the joint. Where leaks take place they are associated with differential deflections between wall panels, and these are worst near corners.

Because bentonite cannot penetrate impervious clays, filter cake does not form in such ground conditions, hence the suspension is totally displaced by concrete and a perfect soil–concrete contact is obtained. Conversely in pervious sand or gravel, all the free bentonite is displaced but the advancing concrete cannot displace the soil impregnated with bentonite. Furthermore, the harder bentonite filter cake is not removed, nor any soft filter cake that occurs in cavities.

For practical purposes concrete used for diaphragm walls can be regarded as impermeable. Because of the humid conditions in which curing takes place, shrinkage cracks due to drying are almost eliminated.

The commonly accepted tolerance on verticality is 1:80. However, the control that can be achieved is governed by the soil, the plant and the early detection of any deviation. Obstructions such as boulders or hard strata, especially if inclined, increase the difficulties appreciably and correction normally involves the use of long, heavy chisels. Millet and Perez (1981) recorded that walls have been built to depths exceeding 122 m with less than 150 mm deviation from the vertical, a tolerance of 1:800. However, the depth limit of diaphragm walls excavated with grab equipment usually ranges between 40 and 50 m since, beyond these depths, such equipment tends to deviate beyond the vertical tolerance and therefore can offer poor guarantee on the quality of the joints and, hence, the continuity of the wall. The use of percussion tools with direct or reverse mud circulation allow this depth limit to be exceeded, but their production rate is low (De Paoli *et al.*, 1989). The slurry trench cutter (see below) is a new type of equipment that can excavate diaphragm walls to more than 100 m and guarantee their continuity.

Goto and Iguro (1989) also described the construction of a deep wall with deviation well within tolerance. The drilling unit of the excavator had an adjustable guide and excavation was done by wing bits (Figure 2.7).

During the excavation of a bentonite trench sections of the soil walls,

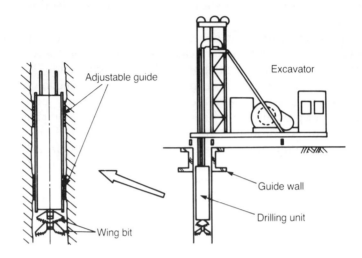

Figure 2.7 Guide wall and excavator (after Goto and Iguro, 1989)

even if only for a short period, have to be free-standing prior to the development of the filter cake. Free-standing height does not present a problem in cohesive soils but, by contrast, in dry coarse-grained soils lacking interlocking particles and surface tension forces due to moisture content, and with no osmotic force between slurry and soil, it is impossible to achieve free-standing height (Veder, 1984).

If excavation work is not carried out properly, then soil movement can occur about the trench. Sliding or caving of the sides with associated slumping of the guide walls and subsidence at the surface may result. Such problems are more likely to occur in coarse- than fine-grained soils. In dry coarse-grained soils they can be avoided by grouting or compacting before excavation begins. If coarse-grained soils are moist, then sufficient time should be allowed for the development of the filter cake, and the rate of excavation should be slower than usual.

It may prove impossible to construct diaphragm walls by normal methods in soils with very low strengths. This is because the internal pressure of the bentonite occupying the trench may be less than the active pressure of the adjacent soil. For example, instability commonly occurs in soft marine clays. In fact, as Sliwinski and Fleming (1975) pointed out, stability is also influenced by the length and shape of the panel and arching effects. They recommended that any soil in which $\phi = 0$ and the cohesion is less than $10\,\mathrm{kN/m^2}$ should be treated cautiously. In such cases panels should be short and uncomplicated.

Another disadvantage concerning the construction of a diaphragm wall is a high water table. This hinders the casting of guide walls. It also reduces the differential between the pressure inside the trench and the active pressure of the adjacent soil. Not only does this have a detrimental effect on the stability conditions but it also adversely influences the formation of filter cake.

If excessive loss of bentonite occurs in pervious strata, especially if the viscosity and gel strength are low, then the formation of filter cake may be poor or non-existent, or may take place only slowly. Viscosity and gel strength may not be adequately developed if the bentonite suspension is not fully hydrated. In fact the rate at which the trench is excavated should take account of the caking properties of the slurry as well as the strength of the soil. A number of measures can be taken to counteract the loss of slurry into adjacent soil, including thickening the slurry by adding further bentonite or cement. Alternatively, the trench can be backfilled with lean concrete, which is excavated shortly after it has started to set. Such measures must be taken immediately otherwise the sides of the trench will begin to collapse as the level of the slurry sinks.

A diaphragm wall was constructed around the site at Sizewell B nuclear power station in Suffolk. It was sunk to a depth of approximately 55 m into the London Clay. Initially, dewatering was considered to lower the water table within the site but reservations were expressed regarding possible settlement that might have affected Sizewell A station. Other considerations taken into account were the prevention of local wells and watercourses drying up and the avoidance of saline pollution of the neighbouring marshes. The plastic concrete cut-off wall has a minimum compressive strength of 1.5 N/mm^2, an *in-situ* permeability of 1×10^{-12} m/s and is 800 mm thick. It contained 50% PFA and the wall can flex with the surrounding ground. Along the foreshore the wall was cast in reinforced concrete.

A new digging unit (Figure 2.8), developed for deep excavation in granular soils or very compact ground, was used (Anon., 1988). The digging machine, suspended from a crawler crane, cuts a vertical slot to the required width and disposes of the excavated material via a hydraulic reverse mud circulation system. Excavation started by trenching to a depth of 3 m with a crane-mounted grab, under bentonite. This provides sufficient depth to submerge the cutters and pump of the digging unit. Before starting excavation with the digging unit a steel guide frame is installed between the guide walls. Each primary panel was 6 m long. Secondary panels in the plastic concrete sections also were 6 m in length but those for the reinforced concrete wall were only 2.4 m. Desanding took place after each panel was excavated. The panel was then concreted by the tremie method. The secondary panels were cut into the concrete of the adjacent primary panels, thus eliminating the need for tube-joints or end-stops to form

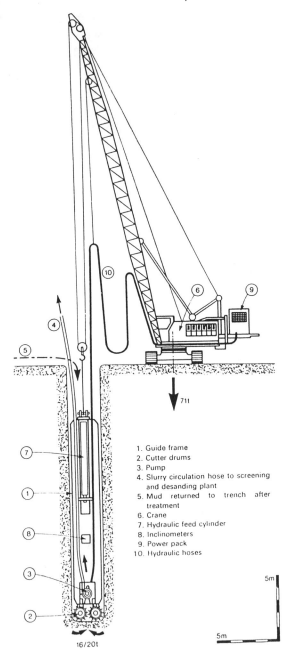

1. Guide frame
2. Cutter drums
3. Pump
4. Slurry circulation hose to screening and desanding plant
5. Mud returned to trench after treatment
6. Crane
7. Hydraulic feed cylinder
8. Inclinometers
9. Power pack
10. Hydraulic hoses

Figure 2.8   Diagrammatic outline of the Soletanche Hydrofraise rig

a watertight seal. It could take 7 days for the plastic concrete to gain enough strength to allow excavation of secondary panels. The required strength was gained much more quickly in the reinforced concrete panels.

Hansmire *et al.* (1989) monitored lateral movement and settlement of a concrete diaphragm wall and found that the movements were small, only a few millimetres. Tie-backs were used as a temporary support for the wall until the associated structure was completed. Their monitoring programme allowed Hansmire *et al.* to make several recommendations for design. One was that if tie-backs are used, their long-term benefit to the final structure should be considered in relation to reducing reinforcement of long spans. Another recommendation was that, in relatively stiff soil in which the strain is kept to well below that where creep movements could develop, the lateral loads assumed in design probably will not be realized. On the other hand, where soils are very soft, long-term lateral loads should be considered.

2.5 THE CHARACTER OF BENTONITE SLURRY

Bentonite slurries, which are used to give support to excavations, can vary widely in their properties; however, according to Hutchinson *et al.* (1975) they must:

1. support the excavation by exerting hydrostatic pressure on its walls;
2. remain in the excavation and not flow into the soil;
3. suspend detritus to avoid sludgy layers building up at the base of the excavation;
4. be easily pumped;
5. be capable of clean displacement by concrete, with no subsequent interference with the bond between reinforcement and set concrete;
6. be able to undergo screening or hydrocycloning to remove detritus and enable recycling.

Usually, the first three items require thick dense slurries, while the latter three are best achieved using very fluid slurries. As a consequence, there is a conflict of requirements which has to be resolved in order that an acceptable specification regarding slurry properties can be developed.

One of the basic functions of a bentonite slurry is to form a filter cake layer on the exposed soil as the trench is excavated. The filter cake only develops if the fluid pressure in the trench exceeds any external groundwater pressure and if the permeability of the soil is within certain limits (Sliwinski and Fleming, 1975).

The bentonite slurry is prepared by adding a measured quantity of the powder to clean fresh water. It is well mixed and allowed to hydrate before it is introduced to the trench. Mixing and hydration are necessary if the slurry is to form a good filter cake. The time taken for hydration, after which the bentonite suspension has attained the desired rheological properties, depends on the method of mixing. For instance, a strong shearing action leads to rapid hydration and good performance.

If a bentonite slurry is to exert a stability pressure on a trench with permeable walls, then it must seal the surface with which it comes into contact. This prevents slurry penetrating into the soil.

Surface filtration occurs when a filter cake starts to build up by hydrated particles of bentonite bridging across the entrances to pores in the soil, without penetrating the soil to any great extent. Water continues to percolate from the slurry through the filter cake into the soil during and after the filter cake has formed.

Deep filtration occurs when a slurry slowly penetrates soil, blocking the pores and developing a filter cake within them. In such instances the seal may extend several centimetres into the soil. The concentration of bentonite in the filter cake, in surface as well as deep filtration, is greater than in the slurry (typically 15% for a slurry containing 5% bentonite).

In this context, determination of the swelling period – that is, the time that elapses between the end of mixing and the introduction of the slurry into the trench – is important (Veder, 1984). For example, when used in highly permeable soils the swelling period should be short to enable the suspension to continue to swell in the trench, thereby preventing deep penetration into the soil with accompanying significant loss of bentonite.

The concentration of bentonite in a slurry has to take into account the often conflicting requirements of trench support, prevention of losses to adjacent strata, reduction of drag and flotation of the grab during digging and the need to prevent too much soil being suspended in the slurry. Normally the concentration of bentonite in slurry varies from about 3% in stiff clays to about 8–10% in dry open ground. For instance, a 4–6% concentration is retained by soil with a permeability up to about 10^{-5} or 10^{-6} m/s. If the permeability of the soil exceeds this, then a denser suspension, ranging up to about 12%, may prove necessary. In exceptional cases, where even this dense fluid is not retained, a number of additives may be tried. However, none of these methods is likely to be very effective if permeability exceeds about 0.5^{-3} m/s. In open ground ($k = 0.01$ m/s), the sealing mechanism can be changed by including small quantities of fine sand in a slurry, with a resultant significant lowering in the initial loss of fluid. Bentonite entering the trench should contain less than 1% fine sand to give a good performance. As far as the formation of a filter cake is concerned, the presence of fine sand can counteract, to a certain extent, incomplete hydration.

The loss of slurry to open dry ground can also be avoided by increasing the viscosity by adding cement (up to about 17 kg per cubic metre of slurry) or by the addition of sawdust, cellulose flakes or shredded clay.

The presence of water in the ground adjacent to a trench can impede the penetration of the surrounding strata. Consequently, slurry concentrations may have to be reduced to a level at which they provide sufficient support to the sides only, usually to between 6 and 8% bentonite concentration. Salty and limey water cause slurry to thicken and, in extreme cases, may lead to bentonite settling out of suspension. In practice, however, these have not proved serious problems. They can be taken care of by adding a phosphate–water mix (up to 1% phosphate) to thin the slurry to its original viscosity. Ground consisting of chalk can give trouble by reducing the effective concentration of bentonite during excavation, thereby diluting the slurry. Peaty water also has the effect of thinning the slurry. In such cases it may be necessary to increase the bentonite concentration to restore the viscosity. In fact, the presence of organic compounds can render the method impracticable.

The pH value is particularly significant where the chemistry of the soil excavated or of the groundwater can give rise to a dramatic change in the pH value of the bentonite slurry. Millet and Perez (1981) proposed that the most desirable range of pH value of slurry is of the order of 6.5–10. If the pH value exceeds 10.5, the clay particles in the slurry will have a tendency to flocculate and settle out. If this happens, then a deflocculating agent has to be added to ensure continued effectiveness of the slurry.

From the point of view of wall stabilization, the higher the density of the slurry the better. The density of freshly prepared slurry varies from about 1.02 t/m^3 for a mix with 4% bentonite to 1.05 t/m^3 for a mix with an 8% bentonite concentration. According to Hutchinson *et al.* (1975), the minimum allowable density should be 1.034 t/m^3. In other words, it is slightly above that of the groundwater. Millet and Perez (1981) maintained that specified maximum density typically is of the order 1.04 t/m^3. This is to ensure that the tremie process is not impeded. In fact, if slurry is to be displaced by tremied concrete its density should not exceed 1.3 t/m^3 and its plastic viscosity should not be above 20 cP* (i.e. 2×10^{-2} N s/m^2; equivalent to a maximum bentonite concentration of about 15%). Fine sand is added to increase the density of a slurry. However, the amount should be limited, again so that the subsequent tremie concreting operations are not impeded. For example, it is generally specified that the sand content should not exceed 5%.

During excavation detritus inevitably becomes mixed with the slurry. Veder (1984) remarked that the incorporation of about 5% fine particles

---

* 1cP = 0·001 N s/m^2.

in the slurry does not impair its effectiveness but a higher percentage leads to a number of shortcomings. Not only does incorporation increase the average density of the slurry, which makes excavation by the grab in the slurry more difficult, but it can generate slowly settling layers of sand and silt. These, in turn, produce density gradients in the slurry and lead to a build-up of sludge in the bottom of the trench. Unfortunately this means that tremied concrete is unable to displace this material from the base of the excavation. Moreover, such slurry tends to have a high viscosity, with adverse effects on displacement of bentonite by concrete past a reinforcing cage. Indeed incomplete displacement of bentonite by concrete can give rise to the formation of cavities.

If the amount of detritus in the slurry slows down the rate of excavation significantly, then the slurry must be replaced. It certainly must be replaced before the reinforcement is placed or concrete poured. If the bentonite slurry is to be re-used, then it must have sandy or silty material removed.

By far the most common form of contamination is by cement, which occurs to a varying degree during tremieing. Contamination by cement leads to the formation of thick permeable filter cakes, very high fluid losses and high viscosities. Hutchinson *et al.* (1975) showed that, in a standard bentonite slurry, the effects become noticeable at 0.3% cement contamination, which is the equivalent to a slurry with a pH value of about 11.7. They therefore proposed that this figure should be the maximum pH value of a slurry.

2.6 COMPRESSED AIR

Compressed air can be used in tunnels, shafts and caissons to prevent the inflow of water and to stabilize the ground exposed by excavation. In soft clay the pressure of the air provides some support, but in sands the air drives the water back from the exposed face so that capillary attraction between the moist grains helps to stabilize the ground. Obviously the permeability of the ground influences air loss and, therefore, whether or not the method can be used. However, compressed air is a health hazard and its use should be avoided where possible. Another drawback is that in large subaqueous tunnels such as vehicular tunnels it is difficult to select an air pressure that will minimize water flow at the invert, where the water pressure is naturally higher, without causing blow-outs at the soffit.

3

Ground freezing

3.1 INTRODUCTION

Ground freezing involves the artificial lowering of ground temperatures so that pore water is converted into ice, thereby reducing the permeability and increasing the strength of the ground so treated. In this way an ice wall can be formed around an excavation until a permanent structure has been constructed. Frozen walls are frequently used in shaft sinking (Auld, 1985, Klein, 1989) and to a lesser extent in tunnel construction (Harris, 1989; Floess et al., 1989). The design of a frozen wall is basically a problem in heat transfer and, in the absence of moving groundwater, transfer in the ground is by conduction. The thermal conductivity (Figure 3.1) and the heat capacity of the ground governs the rate at which freezing proceeds. Fortunately the values of thermal conductivity for all types of frozen soil fall within quite narrow limits. This is one of the reasons why artificial ground freezing is a versatile technique, being able to deal effectively with a great variety of soil types (Jones, 1981).

Artificial freezing can be employed as an exclusion technique to stop the flow of groundwater into excavations, thus producing dry working conditions. In this context, frozen walls can be used in conjunction with bored pile walls or slurry trenches. The frozen ground provides a clear area that is free of bracing and earth supports within which excavation can occur.

Ground freezing is influenced by the geological and hydrogeological conditions at the site, the soil types present, and their thermal properties and water content. Thermal conductivity varies vertically with each change in water content and lithology. It also changes horizontally from the frozen to the unfrozen ground. Obviously the position of the water table and its fluctuation are important, as are the temperature and velocity of groundwater flow. Although complete saturation of the soil

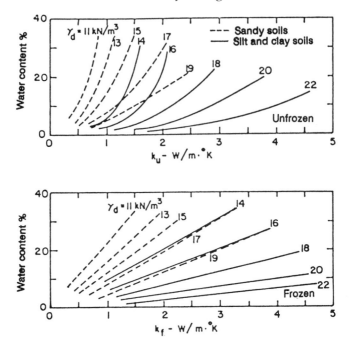

Figure 3.1 Thermal conductivity of soil (from Mitchell, 1981)

is desirable, freezing may be carried out safely when the degree of saturation is as low as 10%. Freezing of pore water in cohesive soils never reaches completion; and attempts at ground freezing may be doomed to failure if there is significant groundwater flow.

The mechanical properties of frozen ground are much more dependent on the time taken in freezing and the ground temperature than on the nature of the soils involved. Consequently, artificial ground freezing is less sensitive to geological conditions than other methods of ground treatment and may be used in any moist soil, irrespective of its structure, grain size or porosity (Jessberger, 1985). Weak, running or bouldery ground, and ground in which the pore-water pressures are high, lend themselves to freezing. Soil profiles, too heterogeneous to grout predictably, are also amenable to freezing. However, in heterogeneous soils the frozen zone tends to be irregular in shape. For a certain refrigerant temperature the frozen ground is usually thinner in silts, clays and organic soils than in sands and gravels (Figure 3.2). As these commonly are the weaker strata, design may be dictated by the former three soil types. When clay soils are frozen, vertical cracks may form in the soil between the frozen columns and the soil may bulge near

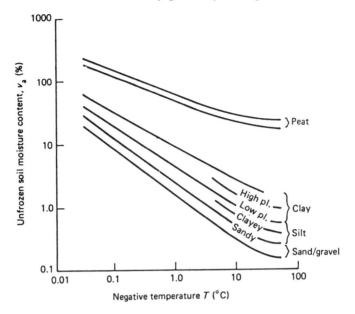

Figure 3.2 Content of unfrozen water in frozen soil (after Frivik and Thorbergsen, 1981)

the zone being frozen. The fibrous structure of organic soils tends to be altered permanently by freezing.

3.2 METHODS OF GROUND FREEZING

Artificial freezing of ground is usually carried out in two stages, referred to as the *active* and *passive* stages of freezing. Active freezing involves freezing the ground to form the ice wall, while passive freezing is that required to maintain the established thickness of the ice wall against thawing. The refrigeration plant has to operate at a much higher capacity during the active than the passive stage of freezing.

The primary plant with pumped loop secondary coolant is the system most frequently used to freeze ground (Figure 3.3). An ammonia or freon refrigeration plant provides the primary source of refrigeration. The system is worked by either electricity or diesel motor and the condensers may be air or water cooled. These plants have a wide range of capacities. Two stages are required to produce temperatures below −25°C. Several different types of coolant have been used with this system; however, the most common type of refrigerant is calcium chloride brine. Other brines have been made from chlorides of sodium, magnesium or lithium. The crystallization point of the chosen brine

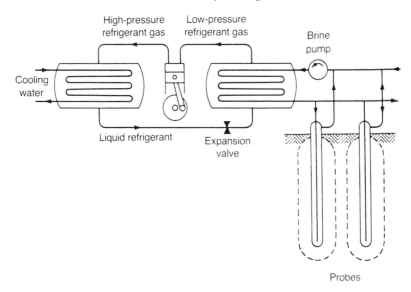

Figure 3.3 Ground freezing with supercooled brine

should be at least 5°C lower than the ultimate temperatures at which ground freezing will proceed. Typical brine temperatures for commercial refrigeration plants are around −20°C to −40°C. A temperature of −5°C can normally be attained throughout the zone to be frozen in reasonable freezing times and freeze-pipe spacings (Sanger, 1968).

Heat is removed from the ground by way of probes placed in boreholes. The installation of the freeze probes represents the most costly, time-consuming and risky part of a ground-freezing project. Each probe consists of an external pipe, closed at the lower end, and containing an open-ended inner tube of slightly shorter length (Figure 3.4). Cooling fluid is normally introduced through this inner tube. The coolant emanating from the bottom of the inner pipe is warmed by the abstraction of heat from the ground as it moves upwards. Large quantities of coolant have to be circulated to bring about freezing. The supply and return temperatures of the coolant in the freeze probes are monitored to determine the amount of heat removed and whether areas are being under- or over-cooled. Such situations can sometimes be corrected by varying the flow of brine. The temperature gradient along a freeze probe is greater when the brine is circulated slowly. The refrigeration capacity of each freeze probe can only be increased by increasing the capacity of the entire system. However, the ability to increase the capacity of individual probes is desirable in order to control localized conditions, especially unexpected groundwater flows.

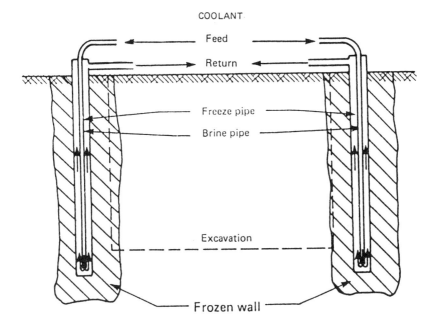

Figure 3.4 Ground freezing layout for an excavation

The growth of ice is more rapid at the base of the freeze probes. This can be overcome by circulating the coolant more quickly or by installing a ring line about 0.3 m below the ground surface immediately inside the perimeter of the freeze pipes, thereby forming a beam of ice.

A warmer brine is discharged from the head of each probe into a collection main, whence it is piped back to the refrigeration plant. There it is pumped through a chiller and is delivered, re-cooled, via a distribution main to the inner tubes of the freeze probes. Such a system, with two compressors, can ultimately lower the brine temperature to −40°C if required.

All piping above the surface in a freezing system must be insulated to minimize heat loss. Insulation is also necessary in large shallow excavations. A temporary cover is desirable for large excavations which may be open for months. However, in a deep excavation of small diameter such as a shaft, the air is cooled appreciably and, as a result, protection of the exposed surface of the frozen soil may be unnecessary.

Expendable refrigerants are very attractive for individual projects which only last a few days or for projects where the cost of delay is high (for instance, for emergency situations where rapid formation of a frozen wall is required or where particularly low temperatures are needed to deal with heavy groundwater flow). Although no refrigeration plant is required,

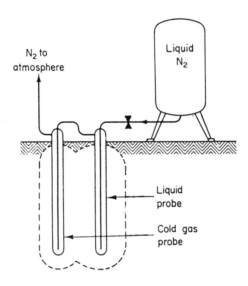

Figure 3.5 Ground freezing with liquid nitrogen

the cost of refrigerant is high. In expendable refrigerant systems the refrigerant is lost to the atmosphere after it has absorbed energy in the ground and vaporized. Liquid nitrogen (LN_2) provides the fastest and thermally most efficient means of freezing the ground (Figure 3.5). It evaporates at a temperature of $-195°C$ at atmospheric pressure and in doing so draws 162.4 kJ/l from its surroundings. About the same amount of heat is needed to raise the cold gas to ambient temperature. In fact such systems can sometimes freeze the ground in a matter of hours. The use of liquid nitrogen also gives rise to fewer ice lenses, to less heaving and to greater strength than freezing with brine. The use of liquid carbon dioxide is thermally less efficient and normally is more difficult to control.

A tanker, from which the liquid nitrogen is injected into the freeze pipes, may be used on small or emergency projects, otherwise one or more storage tanks are needed on site (Gallavresi, 1985). Liquid nitrogen is introduced directly into the circuit, by utilizing the pressures developed in the storage tanks, through a distribution pipe carrying the liquid nitrogen to the freeze probes. These are grouped and connected in series. There is a valve downstream from the terminal freezing pipe through which the nitrogen gas is vented to the atmosphere. The pressure and consumption of liquid nitrogen are regulated at the storage tank in accordance with the desired rate of heat extraction to form the frozen wall.

It is possible to combine brine and liquid nitrogen freezing. In this

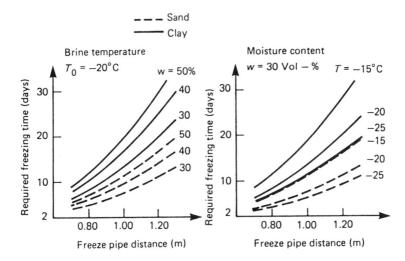

Figure 3.6 Required freezing time versus distance between freezing pipes (from Jessberger, 1985)

way the freezing capacity can be raised to meet peak demands, or liquid nitrogen can be used as a standby in the event of failure. Another possibility involves freezing the ground with liquid nitrogen and then keeping it frozen with cooled brine. Liquid nitrogen can be very quickly re-applied if unforeseen difficulties arise.

Most ground-freezing contractors have evolved their own method of calculating the thickness of a frozen wall. When the thickness of the frozen wall has been decided, the number of freeze holes and their spacing together with the refrigeration capacity is ascertained. The most important factors governing cost are the size and spacing of freeze pipes (Figure 3.6). Estimation of the time required to freeze a radius R by using freeze pipes with radius r_o can be obtained from Figure 3.7. Relative spacings, R', in excess of 15 normally are not used. Shuster (1972) pointed out that without the correct field control of the liquid nitrogen freezing process, the potential time savings shown in Figure 3.7 may not be obtained. The energy requirement needed to freeze soil in kcal/m^2 is approximately 2200 to 2800 times the percentage natural moisture content.

A frozen ground-support system is a massive relatively rigid structure, which is formed generally before excavation commences. Because of the relatively high compressive and low tensile strengths of frozen soils, curved arched walls, particularly circular walls, provide the best structural solutions and should be chosen whenever possible. Although a gravity wall may be constructed in a straight line, the volume of

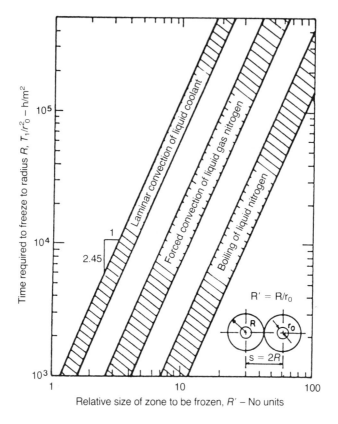

Figure 3.7 Determination of required freezing time (after Shuster, 1972, with permission of ASCE). Indicated bands represent normal range of observed field and laboratory results. However, results with forced convection of LN_2 may vary more widely than indicated due to the variables in control of the freezing process

ground that must be frozen varies from two to more than five times as much as that required for a comparable curved wall.

3.3 HYDROGEOLOGY AND THE NATURE OF FROZEN SOILS

Hydrogeology is one of the most important factors in an evaluation of the feasibility of ground freezing, notably because of the heat contained in moving groundwater. In addition, the position of the water table plays a significant role in ground freezing because of the importance of water content as far as the mechanical properties of the frozen ground are concerned. Some coarse-grained soils that are unsaturated may remain pervious when they are frozen. Hence, in such a situation if the water table rises outside of the excavation after freezing has been completed, water may seep through and cause thermal erosion of the frozen wall. Nonetheless, as mentioned above, ground freezing may take place when the degree of saturation is as low as 10%. Even when below 10%, the moisture content of, for example, sands can be increased by injection of water from the surface (Gonze *et al.*, 1985).

If groundwater is saline or contaminated this can lower the point at which freezing takes place and so reduce the strength of the frozen ground. Where dissolved salts are present in the groundwater, chemical analysis should be made and the temperature at which freezing occurs ascertained. Sea water presents no special problems in this respect since its freezing point is only about 3°C below that of fresh water. However, when a volume of saline water in granular soils has been confined by a perimeter array of freeze probes, inward advance of the frozen wall causes a progressive increase in the salinity of the unfrozen fluid. In narrow excavations the salinity of the water in the unfrozen core is commonly several times that of the original groundwater. Artificial ground freezing has been used in formations containing natural brines derived from evaporitic deposits. In such cases the temperature of the coolant must be much lower than when groundwater is fresh, for example, it may have to be lowered to −20°C or more.

The ground beneath the base of excavation is extremely important. Ideally, the base of a frozen wall should be located in an impervious layer which forms a seal, preventing the incursion of groundwater into the excavation. Where a frozen wall is not founded in an impervious horizon, an excavation has an open base. Consequently, extreme caution must be exercised to minimize water movement beneath the frozen wall and to avoid piping or heaving of the floor.

Movement of water normally leads to an uneven development of an ice wall in the direction of the flow. In addition, if the velocity of groundwater flow is too great, the frozen soil columns will not merge, thereby leaving windows in the wall. Obviously it becomes impossible to freeze soil if the amount of heat conveyed to the freeze zone by flowing groundwater exceeds that being removed by the freezing system.

It is generally accepted that the normal freezing process is not likely to be successful where the flow of water exceeds 1.5 m/day. In such situations special provisions have to be taken either to lower the flow rate or to cope with the excess heat. If groundwater flow is greater than 1.5 m/day but less than 3 m/day, the spacing between freeze pipes should be reduced or a second row of pipes should be placed on the upstream side (Braun *et al.*, 1979). Lower refrigerating temperatures can be used. Calcium chloride can be chilled to −40°C before increasing viscosity has a significant effect on pumping resistance. If colder temperatures are required, then alternative secondary refrigerants or circulation of the primary refrigerant can be used (Harris, 1989). When the flow exceeds 3 m/day, either the groundwater gradient or the permeability of the formation must be reduced. This is brought about by grouting before or during the installation of the freeze pipes, or by intercepting the flow with wells. Generally the latter method provides a more reliable solution. Liquid nitrogen has been used successfully where groundwater flows of more than 50 m/day have been encountered, but this is expensive.

3.4 STRENGTH AND BEHAVIOUR OF FROZEN GROUND

The long-term strength and stress–strain characteristics of frozen ground depend mainly on ice content, the orientation of ice crystals with respect to the direction of applied stress, the temperature of the ice, and duration of loading (Figures 3.8 and 3.9). The frictional resistance of soil particles is also important and is attributable to the same factors which influence the strength of unfrozen soils – that is, particle shape, particle size distribution, density, porosity and soil structure (Sayles, 1989). Friction usually is the dominant factor determining the strength of

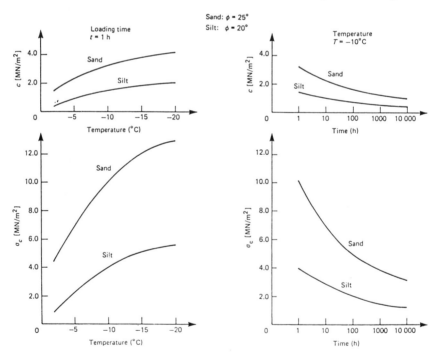

Figure 3.8 Unconfirmed compressive strength and cohesion of frozen sand and silt in relation to temperature and time (after Jessberger, 1985)

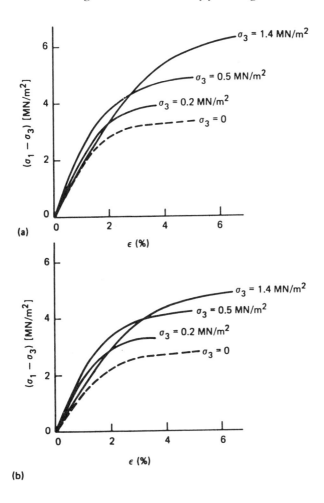

Figure 3.9 Stress-strain relationship of frozen sandy silt at $T = -10°C$: (a) $t = 10$ min; (b) $t = 24$ h (after Jessberger, 1985).

coarse-grained ice-poor soils. The behaviour of ice-rich soils in which few of the grains are in contact is dominated by the characteristics of the ice present. However, the maximum resistance offered by the ice is not necessarily mobilized at the same time or strain as that of the limiting frictional resistance of the soil. The cohesion of frozen soils involves the molecular attraction between particles and the degree of particle cementation, including cementation by ice. Cohesion increases with decreasing temperature but tends to decline with time.

A major part of the resistance to stresses that are rapidly applied is

Ground freezing

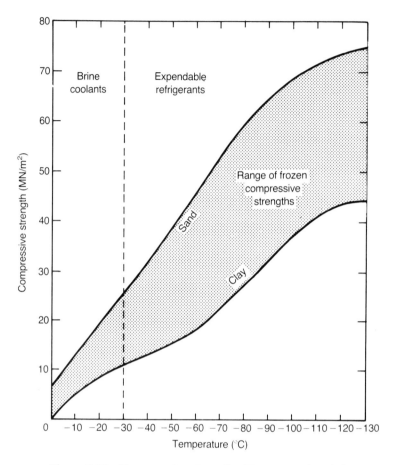

Figure 3.10 Compressive strength of frozen sand and clay

provided by the ice matrix. Under rapid loading the short-term strength may be 5 to 10 times greater than that under sustained stresses, that is, frozen soils are susceptible to large creep strength losses. The compressive strength in short-term tests at low temperatures may be up to 20 MPa.

The more clay material present in a frozen soil, the greater is the quantity of moisture that remains unfrozen. Even so, the unconfined compressive strength of frozen clays undergoes a dramatic increase in strength with decreasing temperature (Figure 3.10). In fact, Lovell (1957) indicated that it appeared to increase exponentially with the relative proportion of moisture frozen. For instance, when tested, silty clay showed that the amount of moisture frozen at −18°C was only 1.25 times that frozen at −5°C, but the increase in compressive strength was more than four-fold. Similar results were found by Kujala (1989). By contrast, the water content of granular soils is almost wholly converted into ice at 0°C. Hence, frozen sands and other soils with relatively large pore spaces exhibit a reasonably high compressive strength only a few degrees below freezing, and there is justification for using this parameter as a design index of their performance in the field, providing a suitable factor of safety is incorporated.

The influence of both time and temperature is very important as far as the stress–strain behaviour of frozen soils is concerned. In particular, stresses and strains in frozen soils induced by externally applied loads are not constant but vary with time. Other factors which influence stress–strain behaviour include the overburden pressure and pore pressure during freezing, the amount of frost penetration, water supply and ice lensing. The time-dependent qualities of strength and viscous behaviour play a decisive role in the design of frozen soil.

Stress concentrations, which occur at the grain contacts when a frozen soil is loaded, bring about pressure melting of ice. Hence, differential water surface tensions are produced which mean that unfrozen water migrates to regions of lower stress, where it freezes. Melting of ice together with water migration disrupt the texture of a frozen soil, leading to plastic deformation of the pore ice and a rearrangement of soil particles. This is responsible for creep (Sayles, 1989). With deformation, the soil becomes more densely packed, giving rise to an increase in strength as a result of an increase in internal friction between grains. There is a concurrent reduction of cohesion and the amount of unfrozen water may increase. If the applied stress is less than the long-term strength of the frozen soil, the weakening process is offset by that of strengthening. By contrast, if the applied stress is greater than the long-term strength, the strengthening process does not compensate for the weakening process and the rate of deformation increases with time. This eventually leads to the failure of the frozen soil mass. Long-term deformations are hundreds of times greater than the initial ones, while the continuous strength is from 5 to 15 times less than the instantaneous strength.

The structured unfrozen water in frozen soil is in equilibrium with the ice and the amount of water present depends on the type of minerals comprising the soil, their specific surface, the temperature, and the salinity of the water. The resistance to stress and deformation of a frozen soil is reduced as the content of unfrozen water increases.

The deformation behaviour of frozen soil is viscoplastic, and both stress and temperature have significant influences on the strain at any time. The creep curves in Figure 3.11 for a frozen silty clay illustrate these effects. The onset of tertiary creep represents the beginning of failure. Creep is temperature and time related for each type of soil. In

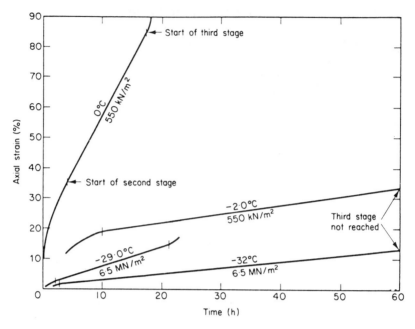

Figure 3.11 Typical creep curves for silty clay at various applied stresses and temperatures (from Bell, 1975)

practice, the evaluation of stability of frozen soil masses, the prediction of creep deformation and the possibility of creep rupture are complex problems. This is because of heterogeneous ground, irregular geometries, and temperature and stress variations throughout the frozen soil mass. However, where the period of unsupported exposure is short and the temperature below freezing point large, the scale of creep relaxation generally is not significant.

Frost heave may occur in some soils when they are frozen, depending on their permeability and *in-situ* confining pressures. Such movement is due to the conversion of pore water to ice on the one hand and to secondary frost expansion which results from migration of pore water to the freeze front on the other (Kujala, 1989). However, if the soil drains at the same rate as the progress made by the freezing front, expansion does not occur. Usually clean free-draining sands and gravels are not susceptible to either type of frost expansion. Significant movements are possible in silts and clays. Mettier (1985) described the reduction of heave from a maximum of 105 mm to 5 mm on certain sections of the Milchbruck tunnel by applying intermittent refrigeration during the maintenance stage.

Further causes of movement associated with artificial ground freezing are a consequence, firstly, of stress relaxation upon excavation and, secondly, of thaw consolidation. If the soil was previously relatively dense, then the amount of consolidation is small. This situation occurs in coarse-grained frozen soils containing very little segregated ice. On the other hand, some degree of segregation of ice is always present in fine-grained soils. For instance, under such conditions the moisture content of frozen silts exceeds significantly the moisture content present in their unfrozen state. As a result, when such ice-rich soils thaw under drained conditions, they undergo large settlements under their own weight. As ice melts and settlement occurs, water is squeezed from the ground by overburden pressures or any applied loads. Excess pore water pressures develop when the rate of ice melt exceeds the water discharge capacity of the soil.

It has generally been assumed that settlement associated with thaw consolidation amounts to 50–90% of heave. However, the two phenomena are independent. For example, 30–80 mm of settlement occurred at Milchbruck tunnel and was more or less the same for sections maintained by cyclic refrigeration as by continuous cooling. Settlement can be controlled by grouting during the thawing stage in order to fill any voids resulting from the freeze–thaw cycle.

4

Drainage techniques

4.1 FILTER DRAINS

Drainage systems can be used to control high-pressure gradients or high pore-water pressures (Cedergren, 1986). Such systems include aggregate filters and synthetic filter materials, as well as pipes and other conduits. Most drainage systems make use of porous filter aggregates to collect the water and conduct it to outlets. Drains must be capable of removing all water that flows into them without allowing an excessive build-up of head. They also should be designed in such a way as to prevent the loss of fines from adjacent soil and, hence, avoid becoming clogged. It is desirable to reduce the loss of head due to flow through a filter to the lowest value compatible with the grain size requirements. Hence effective aggregate filter drains almost invariably require one layer of graded fine aggregate for filtering and a coarse layer of relatively high permeability to remove groundwater.

The following criterion has been used for the design of filters:

$$\frac{D_{15} \text{ (filter)}}{D_{85} \text{ (soil)}} < 4 \text{ to } 5 > \frac{D_{15} \text{ (filter)}}{D_{15} \text{ (soil)}}$$

The ratio of D_{15} of the filter to D_{85} of the soil is termed the piping ratio. In other words, the piping criterion dictates that the D_{15} of the filter material must not be more than 4 or 5 times the D_{85} of the surrounding soil. If the piping criterion is satisfied it is more or less impossible for piping to occur, even under extremely large hydraulic gradients. According to the permeability criterion (on the right-hand side) the D_{15} of the filter must be at least 4 or 5 times the D_{15} of the soil. Although this means that filter layers are several times more permeable than the surrounding soil it does not always guarantee hydraulic conductivity in drains. Ideally the coefficient of permeability should be 10 to 100 times greater than that of the average value of the surrounding soil. The piping criterion should always be used to ensure against piping or clogging when pea gravel and coarse aggregates are used in filter drains. In other words, the filter must be graded, that is, the material should be enclosed by a layer of fine filter material.

Although the above filter criterion is frequently referred to as the Terzaghi criterion, it was actually proposed by Bertram (1940). Since then there have been many studies of filter criteria. For example, Anon. (1955) limited the piping ratio to 5 and used the following criterion:

$$\frac{D_{50} \text{ (filter)}}{D_{50} \text{ (soil)}} \leq 25$$

If filters are placed in plastic clay, then the US Army Engineers allow higher piping ratios. In other words, for such soils the D_{15} size of the filter may be as high as 0.4 mm and the above-mentioned criterion is disregarded. This relaxation in criteria for protecting plastic clays allows the use of a one-stage filter material. However, the filter must be well graded to ensure its non-segregation and therefore a coefficient of uniformity (D_{60}/D_{10}) not exceeding 20 is necessary.

The results of tests carried out by Sherard *et al.* (1984a) on sand and gravel filters showed that there was a very narrow boundary between filter failure and success, when defined by

$$\frac{D_{15} \text{ (filter)}}{D_{85} \text{ (soil)}} = 9 \quad \text{or} \quad D_{85} \text{ (soil)} = 0.11 \, D_{15} \text{ (filter)}$$

Uniformly sized particles of sand with D_{85} smaller than $0.1D_{15}$ (filter) always passed through the voids of the filter, whereas when D_{85} (soil) exceeded $0.12D_{15}$ (filter) they were retained. In well-graded soils (e.g. sandy silts and clays, and well-graded pervious sands) a filter sufficiently fine to catch the D_{85} (soil) size will also catch the finer particles. It was found that for those soils tested, the most widely used filter criterion

$$\frac{D_{15} \text{ (filter)}}{D_{85} \text{ (soil)}} \leq 4 \text{ or } 5$$

was conservative. It can be considered to have a safety factor of about 2. Thus Sherard *et al.* (1984a) concluded that it is still appropriate to use

$$\frac{D_{15} \text{ (filter)}}{D_{85} \text{ (soil)}} \leq 5$$

as the principal filter acceptance criterion. This conclusion applies generally to filters with D_{15} larger than about 1.0 mm. For certain gap-graded and broadly graded coarse soils, usually graded from clay sizes

Table 4.1 Sand or gravelly sand filters and fine-grained soils (from Sherard *et al.*, 1984b)

Soil type	D_{85} particle size of soil (mm)	D_{15} particle size of filter (mm)	Comments on $\dfrac{D_{15}\,(filter)}{D_{85}\,(soil)} \leqslant 5$ criteria
Sandy silts and clays	0.1–0.5	Around 0.5	Satisfactory
Fine-grained silts	0.03–0.1	Less than 0.3	Satisfactory
Clays with some sand content	Greater than 0.1	Around 0.5	Satisfactory
Fine-grained clays	0.03–0.1	Less than 0.5	Reasonable
Very fine-grained clays and silts	Less than 0.02	0.2 or smaller	Satisfactory

to gravels, with D_{85} larger than 2 mm, the soil fines may be able to enter the voids in the filter even if the coarser particles cannot. Sherard *et al.*'s (1984b) recommendations for filters in silts and clays are summarized in Table 4.1.

It frequently has been suggested that a particle size distribution curve of the filter should be approximately the same shape as that of the soil. Sherard *et al.* (1984a) stated that generally this is neither necessary nor desirable).

Sherard *et al.* (1984a) also considered that the average particle size of a sand or gravel filter, D_{50}, does not provide a satisfactory measure of the minimum pore sizes. Therefore, filter criteria using D_{50}, such as

$$\frac{D_{50}\,(\text{filter})}{D_{50}\,(\text{soil})} \leq 25 \text{ and } 12\,\frac{D_{50}\,(\text{filter})}{D_{50}\,(\text{soil})} < 58,$$

are unsatisfactory. They further maintained that filter criteria using D_{15} (soil), for example,

$$\frac{D_{15}\,(\text{filter})}{D_{15}\,(\text{soil})} < 40$$

were even less satisfactory. Hence, they recommended that the use of such filter criteria should be abandoned.

One of the most recent surveys of filter criteria for cohesionless soils has been undertaken by Honjo and Veneziano (1989). Their work showed that the stability of the soil is controlled by the coarser particles, that is, a layer of coarser particles forms at the soil–filter interface which, except for some gap-graded soils, prevents the rest of the particles washing through. They confirmed that the Terzaghi criterion was the most notable predictor of filter performance. Nonetheless, they also found that a D_{95}/D_{75} of 2 or less provides an indication of the capability of the soil to form self-healing layers (i.e. to prevent grains washing through). This they referred to as the self-healing index. They therefore proposed a new criterion involving both the Terzaghi criterion and the self-healing index, which was as follows:

$$\frac{D_{15}\,(\text{filter})}{D_{85}\,(\text{soil})} \leq 5.5 - 0.5\,\frac{D_{95}\,(\text{soil})}{D_{75}\,(\text{soil})} \quad \text{for} \quad \frac{D_{95}}{D_{75}} \leq 7$$

They maintained that this modified criterion provides a uniform safety (i.e. probability of filter failure of about 0.1) over a wide range of soils. In addition, they confirmed what Sherard *et al.* (1984a) maintained, that the

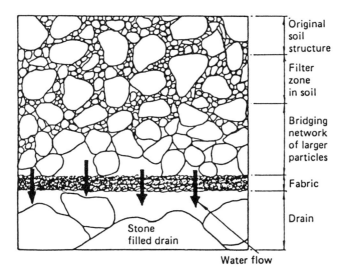

Figure 4.1 Filter drain incorpoating synthetic fabric (geotextile), showing development of a filter zone within the soil

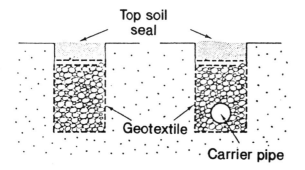

Figure 4.2 Linear drains enclosed with geotextile

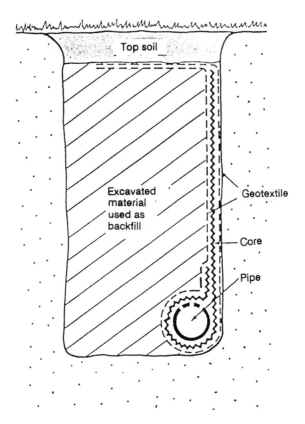

Figure 4.3 Fin drain formed by a composite system of geotextile filter and plastic mesh conducting core

ratio D_{50} (filter)/D_{50} (soil) could not be justified on physical or statistical grounds and that the requirement that the curve of grain size distribution of the filter should be roughly parallel to that of the soil appeared to be unnecessary. However, their single criterion for all soil types would, according to Talbot (1991), result in filters which are too coarse for certain broadly graded soils and soils with significant sand and gravel fractions. Hence, the filter criteria advanced by Sherard *et al.* (1984a and b), which are based on the Terzaghi criteria, are more reliable.

Vaughan and Soares (1982) showed that a relationship exists between the size of particles retained by a filter and its permeability. Hence they suggested that the permeability of a granular filter represented a better way of quantifying particle retention than its grading.

The performance of a filter drain incorporating synthetic fabrics (Figure 4.1) is governed, on the one hand, by the properties of the soil (particle size distribution, shape and packing) in which it is incor-

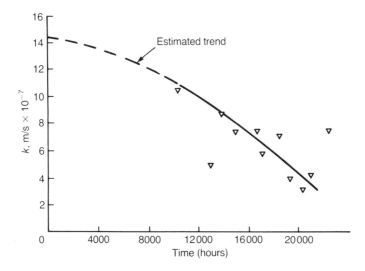

Figure 4.4 Permeability reduction of geotextile installed in the field (after Van der Merwe and Horak, 1989)

porated, and the hydraulic flow conditions (undirectional steady state, transient laminar or turbulent flow, dynamic/reversible flow, and angle of flow with respect to the plane of the fabric); and, on the other, by the properties of the fabric (pore size distribution, permeability, thickness, and variations of these with time and structural loading). Fabrics have been used to form various types of drains such as linear drains where fabric encapsulates granular material in a trench (Figure 4.2); drainage blankets beneath fills; or fin drains (Figure 4.3). In the latter case the geotextile consists of a conducting core enclosed between two layers of fabric (Anon, 1984). Van der Merwe and Horak (1989) have shown that the permeability of a geotextile used as a filter does reduce with time (Figure 4.4).

When pipes are used in drainage systems, the slots in the pipes should be small enough to retain soil particles thereby preventing their movement into the pipes. Nonetheless, there should be enough slots to allow free movement of water into the pipes. Obviously the filter materials around the pipes must have gradations that are compatible with the sizes of the slots or holes. The following criterion is frequently used

$$\frac{D_{85} \text{ (filter)}}{\text{Hole width or diameter}} > 1 \text{ or } 2$$

Drainage techniques

4.2 DRAINAGE OF SLOPES

Drainage is the most generally applicable method of improving the stability of slopes or for the corrective treatment of slides, regardless of type, since it reduces the effectiveness of one of the principal causes of instability, namely, excess pore-water pressure. The distribution of groundwater within the slope must be investigated, as must the most likely zone of failure so that the extent of the groundmass which requires drainage treatment can be defined.

Surface run-off should not be allowed to flow unrestrained over a slope. This is usually prevented by the installation of a drainage ditch at the top of an excavated slope to collect the water draining from above. The ditch should be lined to prevent erosion. It may be filled with cobble aggregate. Herringbone ditch drainage is usually employed to convey water from the surface of a slope. These drainage ditches lead into an interceptor drain at the foot of the slope (Figure 4.5).

The interception of groundwater before it reaches an excavation is easier and more effective than subsequent attempts to remove water once it has percolated into suspect layers. Indeed, as much groundwater flow as possible should be intercepted before excavation commences. For instance, deep interceptor drains, placed at a high level on a site,

Figure 4.5 Construction of surface drainage ditches filled with coarse granular material leading into a lined interceptor drain which, in turn, leads into a sink. The material being drained is till, near Loch Lomond, Scotland

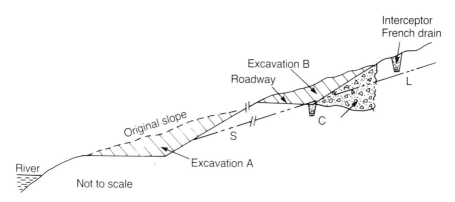

Figure 4.6 Excavation A was made to provide a level recreational area for a school. The ground, being near the bank of a large river in its lower reaches, consisted chiefly of alluvium. Movement started on slip line SL but was unnoticed, for it was seasonal and not very active. Later, a road was to be constructed for access to the school; this work (excavation B) was found to be impossible until the hillside was stabilized. Counterfort drains, C, were constructed about every 3 m over the length of the moving section. The depth of the slip had been found to be about 2.5 m. It eventually became possible to build the roadway

prevent the development of high pore-water pressures near the face of an excavation by stopping the penetration of water.

Water may be prevented from reaching a zone of potential instability by cut-off, thereby allowing natural drainage to occur within the slope. Cut-offs may take the form of a trench backfilled with asphalt or concrete, sheet piling, a grout curtain or a well curtain, whereby water is pumped from a row of vertical wells. Such barriers may be considered where there is a likelihood of internal erosion of soft material taking place due to the increased flow of water attributable to drainage measures.

Support and drainage may be afforded by counterfort drains, where an excavation is made in sidelong ground, likely to undergo shallow parallel slides (Figure 4.6). Deep trenches are cut into the slope, lined with filter fabrics, and filled with granular filter material. Counterfort drains must extend beneath the potential failure zone, otherwise they merely add unwelcome weight to the slipping mass.

Horizontal drains have been used to stabilize clay slopes. They remove the water by gravity flow. Consequently they have to be installed after the area has been excavated a metre or so below the elevation at which the drains will outcrop on the slope. The effectiveness of the drains is governed by their diameter and spacing (the larger the diameter or the lower the spacing, the greater the increase in slope

Figure 4.7 Internal drainage gallery in restored slope, Aberfan area, South Wales

stability), as well as their location in relation to a potential critical slip zone. In other words, the amount of improvement depends on how closely the drains are positioned in relation to the critical zone but there is no additional benefit in extending the length of the drains beyond where this zone intersects the top of the slope.

Successful use of subsurface drainage depends on tapping the source of water, on the presence of permeable material which aids free drainage, on the location of the drain on relatively unyielding material to ensure continuous operation (flexible, PVC drains are now frequently used) and on the installation of a filter to minimize silting in the drainage channel. Drainage galleries are costly to construct and in slipped areas may experience caving. They should be backfilled with stone to ensure their drainage capacity if they are likely to become partially deformed by subsequent movements. Galleries are indispensable in the case of large slipped masses (Figure 4.7), in some instances drainage has been carried out over lengths of 200 m or more. Drill holes with perforated pipes are much cheaper than galleries and are satisfactory over short lengths but it is more difficult to intercept water-bearing layers with them. When individual benches are drained by horizontal holes, the latter should lead into a properly graded interceptor trench, which is lined with impermeable material.

4.3 SAND DRAINS, SANDWICKS AND BAND DRAINS

The theory of consolidation advanced by Terzaghi (1925, 1943) proposed that the time taken for a given soil to reach a certain degree of primary consolidation varies directly with the square of the longest drainage path. In the case of different types of soils the times taken are inversely proportional to permeability. Obviously if the distance water has to travel in a mass of soil is reduced, then consolidation will occur more quickly. This can be achieved by employing drains.

A regular pattern of vertical drains permits a radial as well as a vertical flow of water from the soil. The water from the drains is conveyed into a drainage blanket placed at the surface or to highly permeable layers deeper in the soil. With water escaping radially to the drains, in addition to escaping in the vertical direction, the amount of time taken for consolidation can be reduced to a fraction of that for vertical flow only, the time being governed mainly by the spacing of the drains. Hence it can be concluded that the purpose of the drain is to provide an easier path for the excess pore water to follow as it is squeezed out of a layer of soil during consolidation. In addition to an acceleration in the rate of consolidation, drains also bring about an acceleration in the rate of gain in shear strength and reduction in the lateral transmission of excess pressure.

Sand drains are particularly efficient in stratified soils because of the higher permeability parallel to the bedding (Figure 4.8). Indeed, sand drains have been used in conjunction with dewatering systems to lower the groundwater level in stratified soils that contain alternate layers of pervious and relatively impervious soil. Sand drains conduct water from the more permeable layers to the dewatering screens which are located opposite the less permeable layers. This reduces the cost of dewatering. However, it has been suggested that sand drains probably serve no purpose where design loadings do not exceed pre-consolidation stresses. In such situations the rate of consolidation is more or less the same, whether or not drains are used. Furthermore, drains may not be particularly effective where secondary consolidation is the more important component of settlement, as for example, in peats and organic clays. Nowadays wick or band drains (see below) are more frequently used than sand drains as they are generally more economic.

The coefficient of consolidation for radial flow and vertical compression dominates the design of sand drains and is influenced primarily by the horizontal permeability of the soil and to a somewhat lesser extent (in the sense that it is less variable), by the coefficient of compressibility (Barron, 1948). It also varies according to the type of loading process and

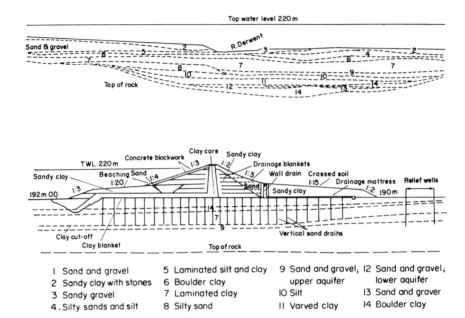

I Sand and gravel	5 Laminated silt and clay	9 Sand and gravel;	12 Sand and gravel;
2 Sandy clay with stones	6 Boulder clay	upper aquifer	lower aquifer
3 Sandy gravel	7 Laminated clay	10 Silt	13 Sand and graver
4. Silty sands and silt	8 Silty sand	11 Varved clay	14 Boulder clay

Figure 4.8 (a) Derwent dam showing the complexity of glacial deposits under the deepest part of the centre-line section; (b) Section through the dam and foundation, showing the horizontal clay blanket linking the clay core with the clay cut-off, the vertical sand drains which hastened the consolidation of the laminated clay, and the relief wells into the upper aquifer (after Ruffles, 1965 *Journal Institution Water Engineers* **19**)

the effective stress level, especially around the preconsolidation pressure. The time, t, taken for consolidation brought about by radial flow to a vertical well, can be obtained from

$$t = T_r \frac{D^2}{c_{vr}} \qquad (4.1a)$$

$$= \frac{T_r D^2 \, a_v \gamma_w}{k_h (1 + e)} \qquad (4.1b)$$

where T_r is the dimensionless time factor for radial consolidation, D is the diameter of the equivalent cylindrical volume of soil being drained by the well, a_v is the coefficient of compressibility ($= \Delta_e/\Delta_p$, i.e. change in void ratio with change in loading), γ_w is the unit weight of water, k_h is the horizontal permeability of the soil, and e is the void ratio. The coefficient of consolidation for radial drainage in a homogeneous soil can be determined from laboratory tests on undisturbed samples and normally will equal or exceed values obtained from laboratory consolidation tests with vertical drainage. Assessment of radial flow is particularly

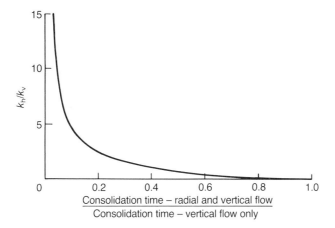

Figure 4.9 Effect of permeability ratio on sand drain efficiency ($H = 3$ m, $D = 3$ m, $d = 0.3$ m, $c_v = 0.096$ mm^2/s) (after Richart, 1959, with permission of ASCE). $H =$ height, $D =$ area of influence, $d =$ diameter of drain, $c_v =$ coefficient of consolidation

important when sand drains are used in laminated soils (Rowe, 1964). However, design values for the coefficient of consolidation for radial flow to sand drains must take into consideration the influence of variable permeability within a formation, especially the influence of any continuous thin layers of sand or silt that can dominate the permeability of a clay soil. The effect of such layers can be estimated from the results of field permeability tests. Usually the ratio of horizontal to vertical permeability for such soils ranges between 5 and 10. The influence of the permeability ratio on the effectiveness of sand drains is shown in Figure 4.9.

The effects of soil disturbance due to the installation of sand drains that can give rise to a smear zone about a drain should be taken into account when assessing the coefficient of consolidation for radial flow (Richart, 1959). The development of a smear zone depends upon how the hole for the sand drain is formed in that mandrels are more likely to give rise to smear than augering. In addition, the density and pore-water pressure of the soil may affect the development of a smear zone. The extent of the smear zone may be quantified in terms of the altered zone so that a smear ratio, SR, can be defined as

$$SR = \frac{\text{radius of smear zone}}{\text{radius of drain}} \tag{4.2}$$

Normally when $SR = 1$ the effect due to smear is of no consequence, whereas when $SR = 1.2$, then the drainage time can increase by up to 200%. Vertical drainage probably will be as effective as sand drains if $SR = 1.5$.

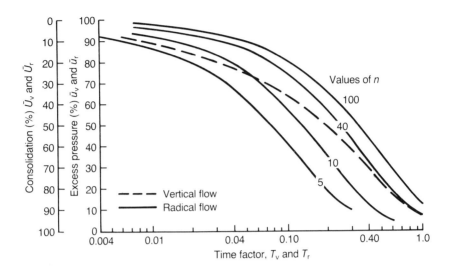

Figure 4.10 Average consolidation rates (after Barron, 1948, with permission of ASCE). The dashed line represents vertical flow in a compressible clay stratum of thickness 2H drained at the top and bottom; the solid lines represent radial flow to axial drain wells in clay cylinders having various values of n

The spacing and pattern of drains are now fairly well standardized. Barron (1948) developed curves (Figure 4.10) which related the average radial and vertical excess pore-water pressures u_r and u_v, and average radial and vertical percentage consolidation U_r and U_v with time factors for radial and vertical flow. Curves are shown for values of n, that is, the ratio of the diameter of the zone of influence, D, to the well diameter, d. In order to determine the well spacing that corresponds to a given value of D, the vertical cylinder surrounding a well is replaced by a vertical prism having an equal cross-sectional area. At the majority of sites, triangular or square patterns with spacings between 1 to 4 m are used. It has been shown that the diameter of the dewatered cylinder varies from 1.05 times the spacing when drains are positioned on a triangular grid to 1.13 times the drain spacing when they are placed on a square grid (Figure 4.11). Spacings of 1.5–2.5 m are the most commonly adopted.

The effectiveness of sand drains is influenced much more by their spacing than by their diameter (Figure 4.12). For example, if the spacing between sand drains is doubled, then this increases the time taken to achieve 90% consolidation by roughly a factor of 6. On the other hand, by reducing the diameter of a sand drain to a twentieth of its size, only increases the time taken for 90% consolidation by a factor of

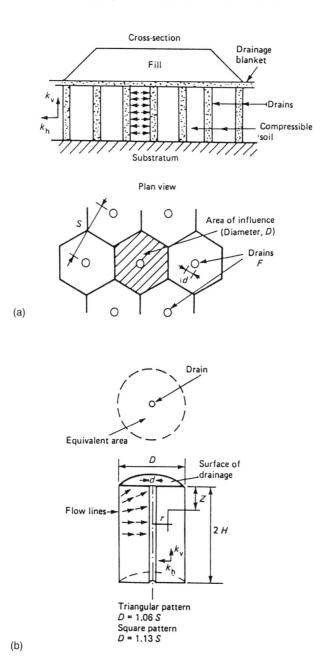

Figure 4.11 (a) Typical vertical drain installation layout; (b) theoretical pattern of drainage through a sand drain

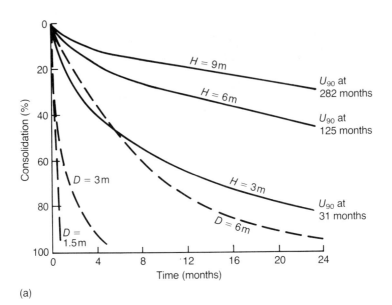

(a)

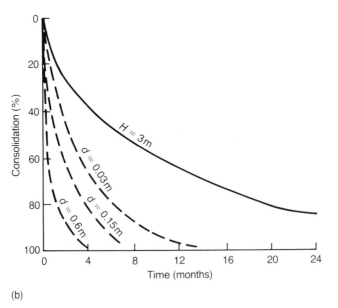

(b)

Figure 4.12 Effect of drain installation and drain spacing and diameter on consolidation time for a clay soil (in (a) $d = 0.3$ m; in both $k_h = 5k_v$ and $c_v = 0.096$ mm^2/s) (after Richart, 1959, with permission of ASCE)

approximately 4. In fact the time for consolidation is proportional to the 2.5 power of the drain spacing.

The depth to which drains are installed is normally equal to the thickness of the soil concerned. For depths up to 20 m vertical drains often prove an economic solution. Beyond 20 m depth, however, the cost of placing drains increases sharply due to the extra effort involved in their installation. For instance, even normally consolidated clays are firm to stiff at depths of 30–40 m. In fact, McGown and Hughes (1981) questioned the value of treating such depths since most settlement and likelihood of shear failure occurs at shallow depths.

Sand drains generally are constructed by inserting casing into the ground and then placing sand in the hole, under air pressure as the casing is withdrawn. The casing may be driven or jet placed and the soil displaced by using a closed-end mandrel, jetted from the hole (Figure 4.13), or removed by augers or rotary-drilling methods.

The installation of sand drains means that consolidation occurs at a faster rate near the sand drains, thus causing a larger amount of surface

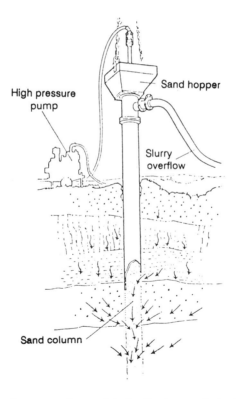

Figure 4.13 Emplacement of a sand drain. Equipment is designed to jet through varying subsoils and leave a vertical column of graded media in the ground

settlement in that region. Such differential settlement can be responsible for a redistribution of surface loading. In fact sand drains of large diameter in a foundation may act as piles in soft soils, attracting load and somewhat relieving stress in the surrounding compressible deposit.

Generally, a drainage blanket of granular material is laid over the area where vertical drains are installed. A considerable quantity of water may be discharged into the drainage blanket, especially during the early stages of construction, if a large volume of soil is being drained. The amount of water that can be discharged into the drainage blanket from a given layout of drains can be derived for any particular degree of consolidation. In this way the effectiveness of the drainage blanket can be assessed and, if necessary, its design can be adjusted accordingly.

As a reduction in the diameter of a sand drain reduces the efficiency of drainage by a slight extent, and as it has proved difficult to construct sand drains with diameters smaller than 250 mm, the sandwick was introduced (Dastidor *et al.*, 1969). Sandwicks are formed by pneumatically filling stockings made of woven-bonded polypropylene with graded sand. The most convenient diameter for sandwicks is around 65 mm. They are installed in holes of small diameter and have proved effective when used to consolidate suspect alluvial deposits.

Sandwicks can be emplaced by a variety of methods depending on the type of soil involved (Robinson and Eivemark, 1985). These include solid or hollow stem flight auger rotary drilling, rotary wash boring, jetting, and driven and vibrated casing. They can be placed to greater depths than sand drains and the fabric stocking ensures continuity at the time of placement.

Although a larger number of sandwicks are needed to bring about the same rate of consolidation as small sand drains, sandwicks have several advantages. The fabric stocking means that a sandwick is continuous, and flexible, and so can adapt to vertical settlement or lateral deformation. It acts as a filter so reducing the possibility of clogging. The likelihood of intercepting lenses of dubious material in heterogeneous soils is far greater if sandwicks with smaller spacings are installed rather than more widely spaced sand drains.

The efficiency of a drain largely depends on its cross-sectional circumference rather than on its cross-sectional area. Hence band-shaped drains are more effective than drains having a circular cross-section. Since the early 1970s man-made fabrics – notably polyethylene, PVC, polypropylene and polyesters – have been used increasingly to produce band drains.

Several types of flat band drains are available (Table 4.2). Most consist of a flat plastic core containing drainage channels surrounded by a thin filter layer (Figure 4.14). Engineering filter fabrics are now more commonly used than treated paper wrappings. Lee *et al.* (1989) referred to

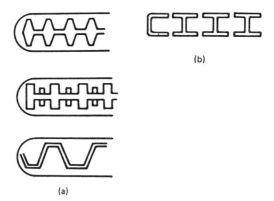

Figure 4.14 Typical cross-sections of plastic band drains: (a) with filter sleeve; (b) without filter sleeve

the use of fibre drains made from jute and coir which have functioned for periods of 12–15 months. In countries where these materials are abundant, the principal advantage of fibre drains is their low cost. Although the permeabilities of the wrappings, especially of paper, may be as low as 10^{-8} m/s, they are thin and offer little resistance to flow. The permeability of the filter should not be higher than that required for the discharge capacity (Hansbo, 1979). This means that filter permeability of existing prefabricated drains need not exceed 0.01–0.1 m/y, irrespective of whether drains are placed in clay or silt. The mechanical characteristics of band drains are important. In particular, the tensile strength usually ranges from 0.5 to 2 kN and elongation at yielding is between 6 and 10%.

The discharge capacity of band-shaped drains varies considerably, depending upon the type of drain used. It also is a function of the effective lateral earth pressure against the drain sleeve. In most cases the filter is partially squeezed into the channel system of the core due to the pressure of the surrounding soil. Accordingly, the discharge is lowered as a result of the reduction in cross-sectional area. Furthermore, the discharge capacity is reduced by 'ageing' once the band drain is installed (Figure 4.15). Other factors that may reduce the effectiveness of band drains include fines entering the channel system and buckling of the drain under large vertical strains. According to Bergado *et al.* (1991), smear effects due to installation may significantly reduce the effectiveness of band drains. If a drain functions correctly during the very early stages of consolidation, the excess pore water contains a small amount of soil particles which may clog the drain. Those drains that do not have open channels in their central area are especially susceptible.

Band drains generally are installed by displacement methods, usually

Table 4.2 Main characteristics of some common types of band drains

Data	Units	Geodrain	Geodrain KB	Desol
Drain body material		Polyethylene	Polyethylene LD	Polyolefine
Filter sleeve material		Cellulose fibre	Polyester Cellulose	Without
Grooves (number)		27×2	27×2	24
Weight (including sleeve)	g/m	168	152	44
Width	mm	95	95	95
Thickness	mm	4	4	2
Free surface area	mm^2	220	220	150
Lateral water permeability	m/s	10^{-6}	10^{-6}	10^{-3}
Length in each roll	m	150	150	400
Roll weight	kg	28	28	20
40 ft container	m	59 400	59 400	160 000

by a lance (Figure 4.16). Speeds of installation are generally of the order of 0.3–0.6 m/s. Auger and rotary wash-boring methods usually are unsuitable, but water jetting within an open casing may be used in very soft soils. Experience indicates that drains spaced at 1–2.5 m centres and extending to depths of between 10 and 20 m give the highest production rates for most methods of installation. In practice, drain spacing is rarely

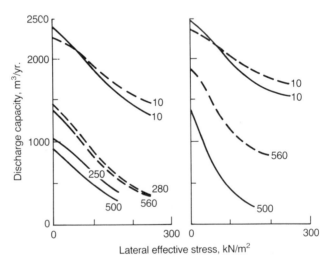

Figure 4.15 Results of discharge capacity tests carried out on geodrains that have been pulled up after having been in peat (left) and slime (right) for different lengths of time (number of days in soil given in figure). Results obtained with filter sleeves of paper shown with full lines and with filter sleeves of synthetics with broken lines (after Hansbo, 1987)

Table 4.2 con'd

Alidrain	Mebradrain MD 7007	Colbond CX 1000	PVC
Polyethylene	100% polypropylene	Coarse polyester monofilament	PVC
Polyester	Polypropylene	Polyester	Without
32×2	38	48	14
180	92	110	100
100	100	100 (150–300 possible)	100
5	3	5	1.8
180	180	220	56
10^{-5}	10^{-3}	10^{-3}	10^{-5}
100−150	250	200	800
	23	25	90
	110 000	80 000	

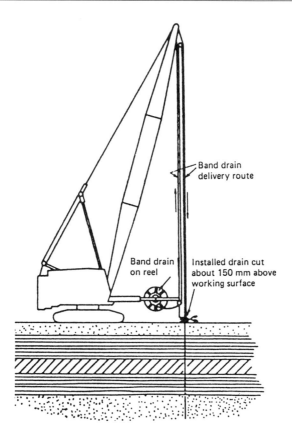

Figure 4.16   Placement of band drains

less than 0.8 m. Band drains have been installed to depths of 40–60 m (Hansbo, 1987).

The effects of smear caused by installing band drains by displacement methods can be reduced by the correct choice of drain filter fabric and the size of the installation lance. A filter fabric initially allows the finer particles of soil, notably those of clay size, to pass through. Hence piping occurs, removing fines from the smear zone, which leads to the formation of a natural graded filter in the soil that may be several millimetres in thickness. This only occurs when the drain fabric has appropriate sizes to allow the development of a natural graded soil filter. Clogging occurs if the pores are too large whereas if they are too small the effect on the smear zone is limited.

4.4 LIME COLUMNS

Cylindrical columns can be formed in clay soils by mixing the clay with unslaked lime (see Chapter 10). This increases the permeability of the columns to between 100 and 1000 times greater than that of the surrounding clay (Broms and Boman, 1979). Hence the lime columns act as vertical drains, as well as reinforcing the soil. One lime column of 500 mm diameter has the same drainage capacity as three 100 mm wide band drains. One of the advantages of lime columns is that their installation creates little disturbance in the surrounding soil, which enhances their performance as a drain.

5

Groundwater lowering

Dewatering is one type of temporary measure taken to lower the groundwater level so that excavation can take place in dry, and therefore more stable, conditions (Bell and Cashman, 1986). However, a change in the position of the water table brought about by drawdown, particularly around a deep, wide excavation may adversely affect the area surrounding the excavation. Of special concern is any settlement likely to occur as a result of withdrawal of groundwater.

5.1 SUMPS

Dewatering brought about by pumping from sumps is the simplest, most widely used – and in many cases the most economical – method of lowering the groundwater level. Pumping from a sump is often the first method tried where an unexpected problem with groundwater is encountered. However, because each sump requires its own pump, the method becomes impractical when more than a few sumps are necessary. A sump must be deep enough to ensure that the entire excavation is drained as a result of pumping. The open sump method is only capable of lowering the water table by up to approximately 8 m. The sump(s) are located below the general base level of the excavation at one or more corners or sides. Continuous pumping of groundwater can remove fines from the surrounding soil if the sump is not surrounded by an adequate graded filter. A ditch or garland drain, which falls towards the sump, is dug around the base of the excavation to convey water from the slopes to the sump. The ditch should be lined or filled with gravel to prevent erosion taking place. Perforated pipe can be placed within the gravel to increase the rate of water removal. Lining the ditch with a filter fabric helps retard the migration of fines.

The pumping capacity needed to establish the initial lowering of the water table is about twice the steady state of equilibrium rate of pumping. Therefore, if double the required steady rate capacity is installed, then all pumps can be used in the initial drawdown period. Subsequently, half the pumps can be shut down and used for emergency standby and maintenance purposes.

5.2 WELLPOINT SYSTEMS

A wellpoint system consists of a number of small wells installed at close centres around an excavation (Figure 5.1). An individual wellpoint usually consists of a gauze screen 0.5–1.2 m long and from 40 to 75 mm in diameter, which surrounds a central jetting riser pipe (Figure 5.2(a)). Various sizes of screen openings are available to correspond with the predominant grain size of the soil. Cashman (1975) recommended that where wellpoints are required to remain in the ground for more than a few weeks, it is more economical to use disposable plastic wellpoints. These consist of a nylon mesh screen surrounding a flexible plastic riser pipe (Figure 5.2(b)). Alternatively, PVC wellpoints, which consist of a plugged tube with an array of fine slots cut through its bottom section, can be used (Figure 5.2(c)). An inner tube is arranged so that the water draining into the wellpoint is drawn upwards from just above the base plug. In both cases the wellpoint and riser are simply cut to various

Figure 5.1   Wellpoint system dewatering an excavation

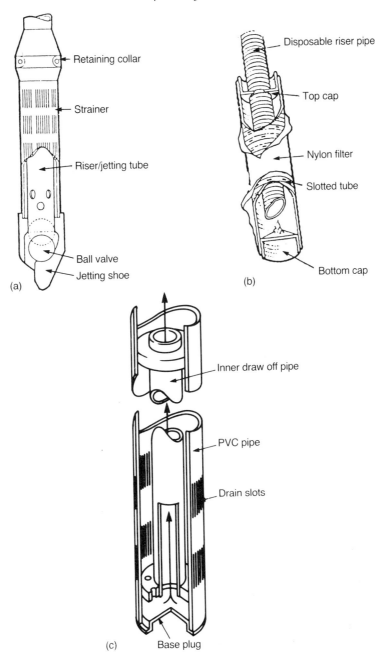

Figure 5.2 (a) A jetting shoe. The latter is attached to the riser pipe. A strainer is located in the recess of the jetting shoe and retained by the retaining collar; (b) Disposable wellpoint and riser pipe; (c) PVC wellpoint

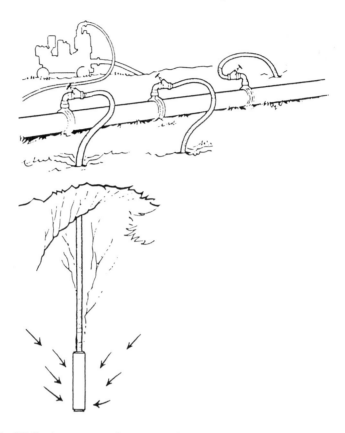

Figure 5.3   Wellpoint system showing wellpoint, riser pipe, header main and pump

lengths to suit ground conditions and depth. At the end of the dewatering period the riser pipe is cut off at or below the ground surface. In fact plastic pipes now tend to be used instead of metal pipes. The individual wellpoints are connected by vertical riser pipes to a header main which is under vacuum from a pumping unit (Figure 5.3). The water flows under gravity to the wellpoint and is drawn by the vacuum to the header main and discharged through the pump.

Because the pumps are on the surface and draw air through the wellpoints, once the groundwater level is lowered they generally have to pass more air than water. This situation can be dealt with by doubling the air capacity of the pumps by using twin-diaphragm attachments instead of a single-diaphragm self-primer (Anon., 1989(a)).

A cone of depression is developed about each wellpoint. The individual cones should intersect in such a way as to lower the water table

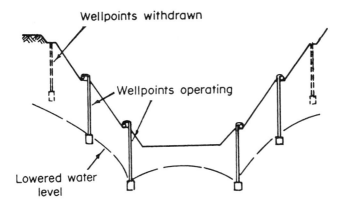

Figure 5.4 Multi-stage wellpoint system

0.5–1.0 m below the base of the excavation concerned. As the wellpoint-ing system draws the water away from the excavation, this means that its slopes are stabilized and therefore can be steeper.

The system is most suitably employed when the water table does not need to be lowered more than about 6 m. Multi-stage systems generally are required beyond this depth because of the limitation of suction lift (Figure 5.4). Nonetheless, improvements in the system have meant that higher vacuums can be maintained, so achieving higher suction lifts and enhancing the ability to dewater fine-grained soils (Figure 5.5).

The main advantage of a wellpoint system is its versatility (Cashman and Haws, 1970). In particular, it can be readily adapted to unexpected groundwater conditions: for instance, wellpoints can be added at

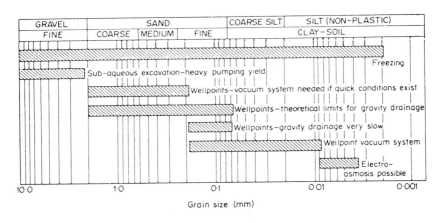

Figure 5.5 Soil types and dewatering systems

offending locations, additional pump capacity can be supplied and wellpoint screens can be raised or lowered with a minimum of trouble.

The most commonly used wellpoint is the self-jetting type, which can be installed rapidly (often in a matter of a few minutes) by jetting water from the nozzle of the wellpoint (Werblin, 1960). On average, around 50 wellpoints can be installed in one day. However, under ideal conditions in clean sand, up to 100 may be put in place. Conversely, with difficult jetting conditions output may drop significantly. Each wellpoint needs about 1 m^3 of water or more to jet it into the ground. The water frequently has to be supplied from a tanker, but once a few wells have been established ground and jetting water can be recycled into a tank on site. Self-jetting wellpoints are suitable for installation in gravels and sands, and can penetrate silts and soft to firm clays. They are suitable for soils that do not yield more than 40–100 l/min, depending on diameter and screen area. For capacities up to 140 l/min per wellpoint, high-capacity wellpoints are required but they are not jetted into place. Large-diameter wellpoints (160–200 mm) are referred to as suction wells and are used where the capacity exceeds 140 l/min.

Occasionally problems arise when jetting occurs in soils containing highly permeable zones of very coarse sand and gravel. The operation can be suspended if the jetting water disappears into such zones. In such cases the wellpoint should be raised a metre or so above the offending zone and held there until turbulence reappears. It then should be lowered slowly towards the coarse layer.

Various drilling methods have been used to install wellpoints when jetting is slow due to the soil conditions. For example, drilling with augers, followed by either plain jetting or installation within a casing is common. If a bed of clay overlies the formation to be dewatered, then it is usually more practical to auger than jet through the clay. A hole-puncher may be used to install wellpoints in soils which offer considerable resistance to penetration such as coarse gravels, cobbles or stiff clays, as well as very permeable soils which are subject to loss of boil (Figure 5.6). After reaching the desired depth the cap is removed and the wellpoint installed before the hole-puncher is extracted. When equipped with a sanding casing, the hole-puncher has been used to jet holes up to 600 mm in diameter and 36 m in depth.

A filter surrounds the wellpoint and riser and increases the effective diameter of the wellpoint, which increases the amount of discharge. The filter column above the screen provides vertical drainage through any thin layers of silt or clay present in sands or gravels. In highly permeable gravels a wellpoint usually does not require a sand filter around it, but if it does, then the wellpoint must be positioned in a cased borehole. The casing is withdrawn after the filter sand is placed.

Testing and evaluating the performance of a dewatering system as it is

Figure 5.6 Hole-puncher and casing

installed and operated are advisable because, owing to normal variations in soil conditions, it is usually impossible during the design stage
to determine with accuracy the permeability of strata, the radius of
influence or distance to the effective source of seepage, and losses of
head in the wellpoint system.

The spacing of wellpoints about an excavation is governed by the
permeability of the soil and the time available to effect drawdown.
Where the aquifer extends 3 m or more beneath the subgrade, the
selection of wellpoint type and spacing is based on the quantity of water
to be removed. The total flow of the system is estimated and divided by
the length of the header main to obtain the flow per unit length. The
wellpoint spacing is then chosen.

Normally wellpoints are spaced between 1 and 4 m apart (Figure 5.7).
Wells probably should be considered if the spacing is greater than 5 m.
Dewatering an unconfined bed of sand or sandy gravel of moderate
permeability above an impermeable bed represents one of the best
situations for the employment of wellpoints (Mansur and Kaufman,
1962). Under special conditions, however, narrower spacings may be
required. For instance, thin layers of clay often occur in water-bearing

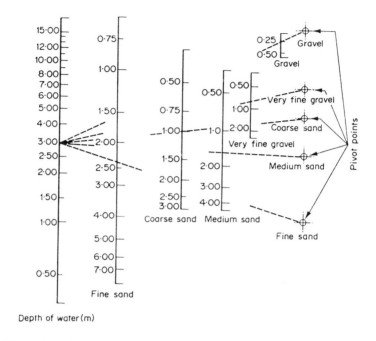

Figure 5.7   A nomogram to determine spacing between wellpoints

Depth of water (m)

sands, and because they are impermeable they give rise to interruptions in the drawdown curves around wellpoints. In such situations well-points commonly are spaced 1 or 2 m apart. Alternatively, holes can be positioned outside the line of wellpoints and filled with coarse sand. Such columns of sand provide a vertical path in the zone of sand being drained and enable water to move more readily towards the wellpoints than towards the sides of the excavation.

Conversely, difficulties can also be brought about by layers of highly permeable material. In this case water tends to by-pass the wellpoints. The problem can be overcome by jetting at close intervals in a line around the excavation, thereby mixing the soils of the different layers.

Wellpoints using vacuum-assisted drainage are effective in silty sands and silts, their spacing usually ranging between 1 and 2 m. A plug of bentonite or cement is placed at the top of the filter column to seal against the entrance of air. Vacuum is applied through the filter column to the soil, thereby accelerating drainage.

In practice the suction lift of a single-stage wellpoint is limited to about 5 m. Under ideal conditions, and using high-vacuum equipment, the depth of lowering can be increased to around 7.5 m. If the total drawdown required is appreciably more than 7m, it is generally necessary to use a multi-stage wellpoint system, or a combination of deep

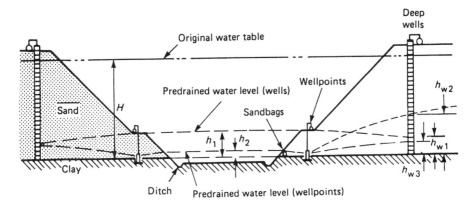

Figure 5.8 A concentric dewatering system using wellpoints and deep wells

wells and a single stage of wellpoints or an ejector system (Figure 5.8). On average each extra stage or tier can lower the water table some 4.5 m.

There is no theoretical limit to the depth of drawdown if a multi-stage system of wellpoints is used. However, the overall width of the excavation has to be increased in order to accommodate this type of layout. Sometimes the upper tiers can be reduced in number or removed after the lower ones have been brought into operation. If an excavation is taken down to the water table before the header main and a pump is installed, then fewer tiers may be used.

The average thickness of the outer part of the slope drained by a multi-stage system is not more than about 4.5 m. Beneath this drained layer, the soil is subjected to seepage pressures produced by percolating water. The stability of the slope should be assessed if the depth of the excavation is more than 12–15 m. This assessment should consider the seepage forces beneath the drained zone.

Where the depth by which the water table has to be lowered is more than 4.5 m but the rate of pumping from each wellpoint is relatively small (that is, less than 45–65 l/min), then the installation of a single-stage system of wellpoints at the top of the excavation or water table, each wellpoint having an ejector pump, may prove more successful than a multi-stage system of wellpoints.

Once the wellpoint system has been established, it is then necessary to dispose of the water. If the system is working properly, it should not be drawing water laden with silt. Most of the fines come through the system in the initial 20 minutes after the start of pumping, and within about 4 hours the water should be clear. Fines will only be drawn off again if pumps are closed down and then restarted.

Horizontal 'wellpoint' drainage is particularly suited to groundwater

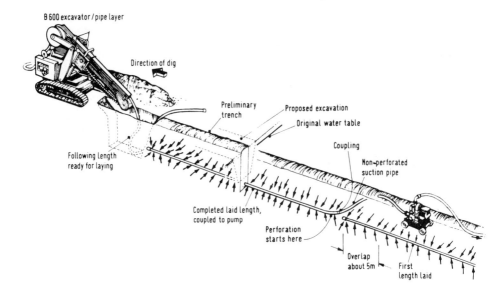

Figure 5.9 Installation of horizontal wellpoint by purpose-built machine (from Somerville, 1986 CIRIA Report 113)

lowering for pipe trenching where fast forward progression is required (Figure 5.9). A long perforated pipe is laid horizontally in the soil at the desired depth by a special trenching machine. This excavates a trench 225 mm wide at depths between 1.8 and 6.0 m and lays thin-walled convoluted perforated plastic pipe of 80–100 mm diameter in the bottom of the trench as the trencher tracks forward. The pipe is wrapped with a mesh of woven nylon or coco-matting. The machine throws the dug spoil back into the excavated trench on top of the perforated pipe. The perforated pipe is cut at given lengths and connected to plain pipe which is brought to the surface at the end of each run. This pipe is then connected to a wellpoint pump. The rapidity with which this system can be installed means that the contractor can progress speedily. This is particularly relevant when laying natural gas pipelines in open country.

5.3 EJECTOR OR EDUCTOR SYSTEMS

An ejector is a jet pump that is used for raising water from a borehole. According to Miller (1988) it is especially well adapted to operating as a group to form a dewatering system. Ejector pumps remain operational in a borehole whether or not they are drawing water. However, they have limited flow capacity so that they are restricted to dewatering less permeable soils.

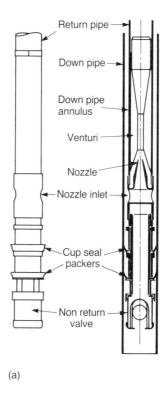

(a)

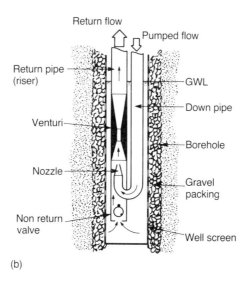

(b)

Figure 5.10 (a) Single-pipe ejector; (b) two-pipe ejector

The simplest type of ejector consists of a downpipe which carries pumped water to an upward-facing nozzle, from which a high velocity jet of water enters the venturi section (Figure 5.10). This jet induces groundwater to enter the venturi and both ground and pumped water flow up the riser pipe to the surface into a tank (Powrie and Roberts, 1990). The induced flow of groundwater is brought about by momentum transfer in the highly turbulent perimeter of the jet (Miller, 1988). Nozzle velocities of around 30 m/s are required to lift water from 20 m depth and 50 m/s to lift from 50 m. Water needed to maintain the operation is recirculated from the tank and the excess disposed of. Cup seals have to be incorporated beneath a single pipe ejector to provide an end-stop to the annular downpipe, thereby preventing pumped flow discharging into the ground.

If the groundwater level is at or below the intake level of the ejector, then air flowing into the throat of the ejector develops a vacuum inside the well, accelerating drainage and reducing pore-water pressures. When air is not available, subatmospheric pressure conditions can develop. This subatmospheric pressure can trigger chemical precipitation, which can cause scaling problems. For example, when groundwater contains more than 50 mg/l of iron its precipitation can block ejectors. Accordingly, an ejector may have to be taken out of service and cleaned occasionally.

Ejector systems are most effective when used to dewater deep excavations made in stratified soils where close spacing is necessary (Figure 5.11). In an ejector system the pumped flow inlets to the individual ejectors are connected to a common pumped flow main. Similarly, the return flow outlets are joined to a common return flow main which discharges into a water storage tank. Junctions between mains and individual ejector well heads have shut-off valves and a union connector so that any ejector can be readily isolated from the rest of the system. Ejector systems are not limited in terms of suction lift as are wellpoints, and have much lower unit costs than wells. Hence, they are used where it is inconvenient or impossible to install a multi-stage wellpoint system since the principal advantage of the ejector system is its ability to operate numerous wellpoints in a single tier to depths in excess of 20 m. Individual ejectors can pump from 20 to 270 l/min (Prugh, 1960). They are installed by jetting within a casing in a similar way to that used for sinking conventional wellpoints in difficult ground, and are surrounded by a sand filter. They can also be installed by water flush rotary drilling, and surrounded by a perforated liner and sand filter. Spacings of 3 m to 15 m are typical. According to Miller (1988), it is not normally economical to space a line of ejectors much closer than 3 m.

The ejector is a self-priming device. It evacuates air from the wellpoint beneath. If one ejector in a system is drawing air, it does not affect the balance of the system, provided air can be vented from the return or

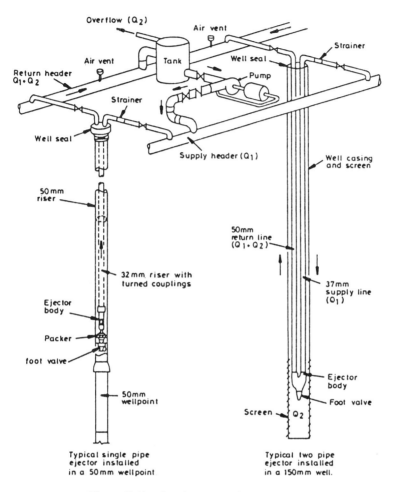

Figure 5.11 An ejector or eductor system

collector main. When the flow of groundwater is less than the design capacity of the ejector and the filter column has been sealed properly, then the unit develops a vacuum in the wellscreen and this is transmitted to the soil. The ability to develop a vacuum in the soil is particularly effective in draining fine-grained soils. Indeed, the maximum practical capacity of an ejector-operated system is 6800–9000 l/min, which usually limits their application to fine sands and silts. If the wells are carefully constructed using appropriate filters and bentonite seals, the effects on fine sands and silts can be dramatic. Only low volumes of water have to be removed from such soils to achieve stabilization so that the low efficiency of ejectors is acceptable.

5.4 BORED WELLS

A deep bored well system consists of three components, namely, the well itself, the pump and the discharge piping (Figure 5.12). The well has an open area into which groundwater flows and a cased section which conducts the water to the surface. In granular soils, slotted metal screens or perforated casings are used to prevent collapse of the walls in the well zone. A graded filter is placed around the screen. During the drilling operation a temporary outer casing is inserted in the drillhole, inside which the permanent inner casing and screen are positioned. The annular space between the two casings is then filled with the filter.

Deep wells have been used for lowering the water table where the soil conditions become more pervious with depth, and have proved par-ticularly worth while where it has been necessary to remove large quantities of water from thick, highly permeable formations. Indeed, deep wells are most effective where the water-bearing soil is clean, uniform and highly permeable such as some sands and gravels. They also are effective where a large excavation is underlain by a relatively impermeable bed, beneath which there is a permeable formation under excessive artesian pressure.

For dewatering purposes deep wells commonly are located at 6.5–65 m centres, depending upon the quantity of water that must be removed, the permeability of the ground, the source of seepage and the amount of submergence available. Such wells usually have a diameter of 150–450 mm, although at times the diameter may exceed 1 m.

The hydraulic efficiency of a deep-well system may be increased by some 10–15% by applying a vacuum line connected to the tops of the wells. The wellhead and annulus must be airtight.

A deep-well system is installed by boring or drilling operations. The selection will depend upon the soil conditions expected and the avail-ability of equipment. Holes extending to 30 m in depth with diameters of up to 600 mm have been excavated by bucket augers. Bucket augers are effective in gravels and sands, and in soft to moderately stiff silts and clays. However, cobbles and boulders present problems and very stiff clays are difficult to penetrate. Excavation by bucket auger can yield a hole of good quality if caving is prevented by using a head of water. A minimum of about 3 m of head is recommended (Powers, 1981). Subsequently, the hole requires little development. In loose permeable sands it may be necessary to mix additive with the boring fluid to form a temporary seal around the hole, thereby preventing caving. Bentonite is not recommended because the filter cake it forms is difficult to remove from the hole, but biodegradable muds are acceptable.

Dewatering wells are occasionally constructed by continuous flight augers. Heavy duty augers can excavate holes with diameters up to 1.2 m.

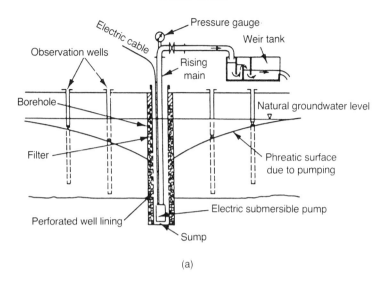

(a)

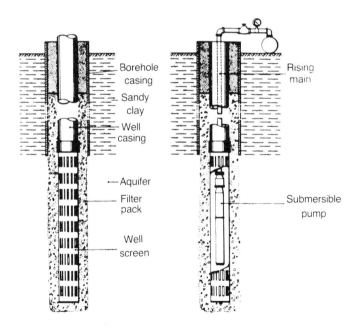

Figure 5.12 (a) A deep-well system; (b) well in more detail

At times heavy slurry is required to stabilize the hole and to aid the retention of sand on the auger as it is being removed. The slurry, together with the mixing of soil and the smearing of the sides of the holes, usually results in a hole of poor quality. Rotary drills, where the cuttings are removed by circulating fluid, are effective for holes of small to moderate diameter. For example, a heavy duty rig can produce a hole 450 mm in diameter.

Reverse circulation rotary drilling is sometimes employed in dewatering programmes. Holes with diameters of 600 mm or more can be sunk. This type of drilling generally results in a cleaner hole than the conventional rotary methods. The method is best suited to loose sands and gravels and soft clays. Stiff clays prove difficult to penetrate and cobbles may also cause difficulty since they tend to become lodged in the drill pipe. A head of water is required to provide support for the sides of a hole when using reverse circulation drilling, a minimum of 3 m from the drilling surface to the water table being recommended. Where the water table is close to ground level the drilling rig can be placed on a platform or, alternatively, a small wellpoint system can be operated while the hole is being bored.

The development of a well is an essential operation in its proper completion. There are three beneficial effects of development:

1. Development corrects any clogging of the water-bearing formation that may have occurred during drilling, and it effects removal of boring slurry.
2. Development tends to increase the porosity of the natural formation in the immediate vicinity of the well and so improves its yield.
3. Development stabilizes the filter media formation stabilizer in the annulus.

In addition, removal of 'fines' during development results in clean water being pumped when the electro-submersible pump is installed (i.e. no fines in suspension), thereby prolonging the useful life of the pump.

There are two basic techniques of well development, namely, surge plunger, either solid or valve type, and air-lift pumping. In general the air-lift pumping technique is to be preferred although it is not effective for relatively shallow wells, that is, less than about 20 m deep.

Depending upon ground conditions, deep wells either may be used exclusively to dewater an excavation or they may be used in conjunction with another pumping system. For instance, wellpoints may be required at the base of the slopes of an excavation in order to intercept any minor seepage that may by-pass the wells. Alternatively, heave or a blow-out may be a possibility where the strata beneath the floor of an excavation are subject to artesian pressure. The artesian pressure can be relieved by the installation of deep wells either around the top of the excavation penetrating into the offending aquifer or at some safe elevation as the excavation is extended downwards. Depending upon the magnitude of the excess pore-water pressure and the soil conditions, relief wells or 'bleeder' wells may be adequate.

Where a site consists of silts or silty sands that are underlain by a bed with higher permeability – for example, of coarse sand – seepage towards the excavation can be intercepted and the water table lowered by a combination of vertical sand drains installed around the top of the excavation and deep wells installed in the coarse sand. The sand drains facilitate drainage of the upper, less permeable soil, while the pore-water pressure in the coarse sand is reduced by pumping from the deep wells (Powers, 1981).

Shallow-bored filter wells are a synthesis of wellpointing and deep wells – some refer to this method as 'jumbo' wellpoints. The installation of the wells is the same as for deep-bored wells with the same facility to ensure controlled filtering, but the wells are pumped by suction pumping either by individual risers to a common suction main or by individual wellpoint pumps at each wellhead. On a congested site, especially in high-permeability soils, this method may be preferred to wellpoints because of the smaller number of risers causing obstruction to the construction operation.

5.5 SETTLEMENT AND GROUNDWATER ABSTRACTION

When a water table has to be lowered in order to construct an excavation, the effect of such an operation on neighbouring structures must be considered. On occasions it may be necessary to keep the groundwater at its original level in a built-up area immediately adjacent to an excavation so that the existing structures are protected from possible damage.

Lowering the water table by abstracting groundwater leads to an increase in effective pressure, which, in turn, can give rise to settlement of the ground surface. Significant settlements can also occur when water is pumped from a confined aquifer overlain by compressible soil. Similarly, abstraction of water from an aquifer containing layers or lenses of compressible soils may cause settlement. The amount of settlement that occurs depends on the thickness of the compressible soils and their compressibility, on the amount by which the phreatic surface is lowered and on the length of time over which pumping takes place (Placzek, 1989). The rate of settlement is governed by the permeability of the compressible soil.

Depending on the site conditions, it may be feasible and desirable to use groundwater recharge to limit the radius of influence of the cone of depression and so reduce potential settlement. In such instances the

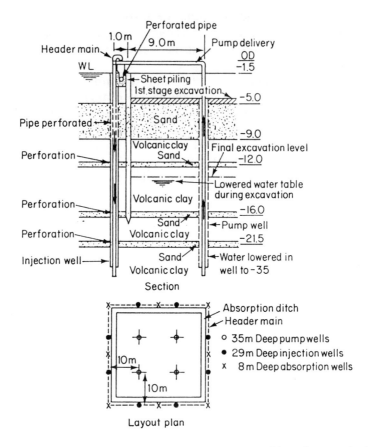

Section

Layout plan

Figure 5.13 A pumping and recharge system to avoid settlement during the excavation of basements (after Zeevaert, 1957)

excavation usually is surrounded by sheet piling with the recharge wells located outside the piling (Figure 5.13). Relief wells within an excavation can be used to reduce piezometric pressures, and thereby overcome the problem of heave.

5.6 ELECTRO-OSMOSIS AND ELECTROCHEMICAL STABILIZATION

Elecro-osmosis was originally developed as a means of dewatering fine-grained soils (Casagrande, 1952; Lo *et al.*, 1991). It consists of passing a direct current from anodes to cathodes positioned at predetermined locations in the soil to be stabilized (Figure 5.14). As the current passes through the soil it causes water to migrate from the anodes towards the cathodes where it is collected and then removed (Figure 5.15). The

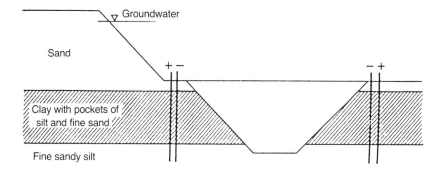

Figure 5.14 Layout for dewatering by electro-osmosis

mechanism responsible for bringing about migration is not completely understood but it appears that absorbed water is removed from clay particles, and that ion transfer through the pore water causes it to flow.

Not only does electro-osmosis reduce the water content of the soil but it also directs seepage forces away from the surface of an excavation as well as developing pore-water tension in the soil (Perry, 1963). Moreover, base exchange occurs in the soil during electro-osmosis, thereby enhancing its strength. In addition, ion migration, electrolysis and chemical reactions occur leading to the formation of new irreversible

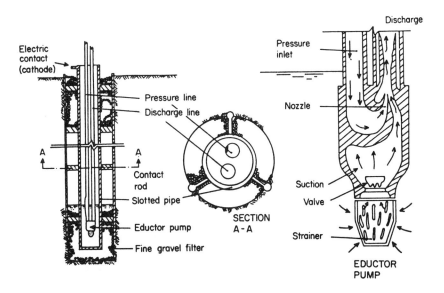

Figure 5.15 A typical cathode well system

compounds. Indeed, chemicals such as calcium chloride and sodium silicate are sometimes added to ensure the growth of cementitious material in the pore space.

Where electro-osmosis has proved successful, it usually has been as a result of using the process to introduce a chemical into the soil, either through anode solution or by direct electrolyte replacement. This improves the stability of the soil either by ionic replacement occurring in the clay mineral content, or by cementitious material being deposited in the pore space. This adaption of electro-osmosis is known as electro-chemical stabilization.

Unlike the coefficient of soil permeability (k), and therefore the rate of hydraulic (laminar) flow, the coefficient of electro-osmotic permeability (k_e) and the rate of electro-osmotic flow are independent of the size of the capillaries. Hence the electro-osmotic permeability is more or less independent of the soil pore size. In other words, it is about the same magnitude for sands, silts and clays, provided that the electro-osmotic potential is about the same for most of the mineral matter in the soil. The electrokinetic potential depends on the concentration and valence of soluble ions in the electrolyte. It is high in a diluted solution, whereas in a concentrated solution it may be reduced to zero. In fact, the potential does not vary much within the ordinary range of concentration of electrolytes present in soils. However, the electrokinetic potential can fall to zero in some situations and the electro-osmotic flow either ceases or may be reversed in direction. For practical purposes, most saturated soils may be assumed to have a coefficient of electro-osmotic per-meability of around 2×10 to 5×10 cm/s for a gradient of 1 volt/cm – for example, Norwegian quick clay = 2×10 cm/s; clayey silt = 5×10 cm/s; and fine sand = 4.1×10.5 cm/s. Electro-osmosis can be used for dewatering silts and clays but the method is not suitable for deposits of sand.

The relative efficiency of electro-osmosis as a dewatering technique may be expressed in terms of the ratio between k_e and k referred to as the coefficient of electro-osmotic effectiveness. It increases with increasing clay fraction in the soil.

Spacing of electrodes depends on the electrical potential available, the potential gradient required for a given soil and groundwater conditions. Spacings between 3.6 and 4.95 m and potentials varying between 30 and 180 volts have proved effective. As far as individual soil types are concerned Zhinkin (1966) proposed a potential gradient of 0.9 volt/cm for sandy loams, 0.7 volt/cm for loams, 0.6 volt/cm for clays, and 0.4 volt/cm for silts. This implies that for a direct current potential of 100 volts, the electrode separation would range from just over 1 m to 2.5 m. For maximum effectiveness, the potential gradient should be in the same direction as the hydraulic gradient. Generally, potential gradients of more than 0.5 volt/cm should not be exceeded for long-term applications because higher gradients lead to energy losses in the form of considerable heating of the ground. However, it might prove advantageous to use potential gradients of 1–2 volt/cm during the first few hours. This gives a much faster build-up of tension in the pore water. Economy of power consumption could perhaps be obtained by intermittent operation.

Negative pore-water pressures develop as water is removed next to the anodes while at the cathodes the pore-water pressures can increase to high positive values. This difference in pore-water pressure produces a hydraulic backflow. It also means that the soil around the cathodes may be softened.

Consolidation may occur as a result of the development of negative pore-water pressure under controlled electro-osmotic drainage (Mitchell and Wan, 1977). This, in turn, leads to changes in the stress–strain and strength characteristics of the soils concerned. The Terzaghi equation of consolidation for homogeneous material also applies to electro-osmotic consolidation where the voltage distribution remains constant with time. Like ordinary consolidation under direct loading, electro-osmotic consolidation is a mass flow process in response to pressure of potential gradients. The rate of consolidation is controlled by the coefficient of consolidation which, in turn, is governed by the hydraulic conductivity and not by the electro-osmotic permeability. For a given compressibility, the amount of consolidation depends on the magnitude of negative pore-water pressure that can be developed at the end of consolidation. At the end of the electrical treatment the effective stress developed during the electro-osmotic consolidation relaxes gradually and some rebound occurs.

Under ideal conditions, and starting with no excess pore-water pressure, at the end of the period of electro-osmotic treatment a maximum pore-water tension occurs at the anodes and decreases linearly to zero at the cathode. Hence, the soil properties vary between anode and cathode. In order to develop a more uniform stress distribution and to increase the average shear strength of the soil, it may prove useful to reverse the roles of the electrodes after a certain period of treatment.

Two factors mitigate against the use of electro-osmosis on a large scale. Firstly, overconsolidated clays tend to be fissured. If the ability of such soil to transmit water along these fissures is similar to the electro-osmotic permeability, then it is unnecessary to use electro-osmotic treatment. Cheaper methods of drainage, such as vertical drains, can be used instead. More seriously, electro-osmosis only operates efficiently in saturated and near-saturated soils. It therefore cannot be used to dewater a soil completely. However, Bjerrum *et al.* (1967) noted that an excessive current density means that the soil around the anodes dries out, thereby increasing the anode–soil resistance so that the process becomes increasingly less effective. Cessation of the process, especially in clay soils, permits reabsorption of water. This occurs at low hydraulic rates but, nevertheless, after a certain period the soil reverts to its original state.

The mechanical properties of clay minerals vary – particularly those belonging to the smectite family – according to the type of cations associated with them. If a clay mineral contains a significant amount of weakly bonded cations such as sodium or lithium, then it has a tendency to absorb and retain large quantities of water. Accordingly, it possesses a low shear strength. By contrast, if the cations have higher bonding strength – such as calcium or magnesium or, more especially, iron or aluminium – the clay material absorbs less water and has a more stable structure.

As cations associated with the clay minerals are exchangeable, the introduction of solutions containing an excess of cations with higher bonding strength gives rise to ion exchange, thereby enhancing the soil properties. Introduction of stabilizing electrolytes into a clay soil is brought about by either solution and dispersion of the anode material through electrolysis or by direct introduction of additives at the anode.

Electrolysis normally involves decomposition of the electrolyte accompanied by solution of anode material. Electrochemical stabilization due to anode solution results from initial base exchange reactions on the surface of the clay minerals, weakly bonded cations being replaced by strongly bonded cations, and the formation of soil-cementing compounds produced by reaction between the electrolyte and dissolved anode solution. Thus a lattice-like structure is built-up around the clay particles. This greatly reduces the amount of settlement that would result from the use of electro-osmosis on its own.

Addition of additives to the anode (in addition to the dissolved electrode material) can increase the base exchange reaction, act as a catalyst in the formation of cementitious material, and improve the quality and quantity of cementitious material. To increase the base exchange reaction, most of the additives used are organic and inorganic compounds of aluminium or calcium, iron normally being present in the form of the electrode material. Of these the most commonly used compound is calcium chloride. This intensifies dissolution of the iron anode by the formation of ferrous chlorides, at the same time producing calcium hydroxide which acts as a cementing agent.

An important disadvantage in the use of additives is the reduction in flow with increase in electrolyte concentration. This may effectively limit stabilization beyond a certain level.

6

Compaction and consolidation techniques

6.1 MECHANICAL COMPACTION OF SOIL

6.1.1 Introduction

Mechanical compaction is used to compact fills and embankments by laying and rolling the soil in thin layers. In other words, it refers to the process by which soil particles are packed more closely due to a reduction in the volume of the void space, resulting from the momentary application of loads such as rolling, tamping or vibration. It involves the expulsion of air from the voids without the moisture content being changed significantly. Hence the degree of saturation is increased. However, all the air cannot be expelled from the soil by compaction so that complete saturation is not achievable. Nevertheless, compaction does lead to a reduced tendency for changes in the moisture content of the soil to occur. The method of compaction used depends upon the soil type, including its grading and moisture content at the time of compaction; the total quantity of material, layer thickness and rate at which it is to be compacted; and the geometry of the proposed earthworks.

With a soil of a given moisture content, increasing compaction results in closer packing of the soil particles and therefore in increased dry density. This continues until the amount of air remaining in the soil is so reduced that further compaction produces no significant change in volume. The soil is stiff and therefore more difficult to compact when the moisture content is low. As the moisture content increases it enhances the interparticle repulsive forces, thus separating the particles causing the soil to soften and become more workable. This gives rise to higher dry densities and lower air contents. As saturation is approached, however, pore-water pressure effects counteract the effectiveness of the compactive effort. Each soil,

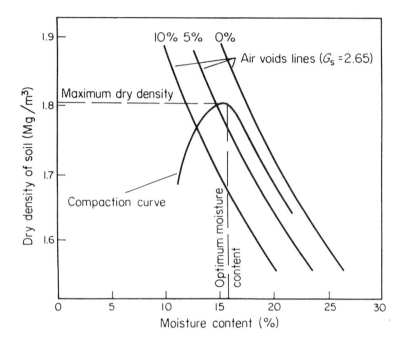

Figure 6.1   A compaction curve indicating maximum dry density and optimum moisture content

therefore, has an optimum moisture content at which the soil has a maximum dry density (Figure 6.1). This optimum moisture content can be determined by the Proctor or standard compaction test (Anon., 1976). Unfortunately, however, this relationship between maximum dry density and optimum moisture content varies with compactive effort and so test results can only indicate how easily a soil can be compacted. Field tests are needed to assess the actual density achieved by compaction on site (Parsons, 1987).

Most specifications for the compaction of cohesive soils require that the dry density that is achieved represents a certain percentage of some laboratory standard. For example, the dry density required in the field can be specified in terms of relative compaction or final air void percentage achieved. The ratio between the maximum dry densities obtained *in situ* and those derived from the standard compaction test is referred to as *relative compaction*. If the dry density is given in terms of the final air percentage, a value of 5–10% usually is stipulated, depending upon the maximum dry density determined from the standard compaction test. In most cases of compaction of cohesive soils, 5% variation in the value specified by either method is allowable, provided that the average value attained is equal to or greater than that specified.

Specifications for the control of compaction of cohesionless soils either require a stated relative density to be achieved or stipulate type of equipment, thickness of layer and number of passes required. Field compaction trials should be carried out in order to ascertain which method of compaction should be used, including the type of equipment. Such tests should take account of the variability in grading of the material used and its moisture content, the thickness of the individual layers to be compacted and the number of passes per layer.

6.1.2 Types of compaction equipment

Sheepsfoot rollers are best suited to compacting cohesive soils in which the moisture content should be approximately the same as the optimum moisture content given by the modified AASHO compaction test. At least 24 passes are required in order to achieve reasonably adequate compaction with a sheepsfoot roller, and the layer of soil to be compacted should not be more than 50 mm thicker than the length of the feet (Table 6.1). The resultant air voids content of soil compacted by sheepsfoot rollers is rather greater than that of soil compacted by smooth-wheel or rubber-tyred rollers.

Rubber-tyred rollers range in weight up to 100 tonnes. The large rollers are usually towed by heavy crawler tractors. Heavy rubber-tyred rollers are not recommended for initial rolling of heavy clay soils but are effective and economical for a wide range of soils from clean sand to silty clay (Table 6.1). The rubber-tyred roller produces a smooth compacted surface which consequently does not provide significant bonding and blending between successive layers. Also, the relatively smooth surface of a rubber tyre can neither aerate a wet soil nor mix water into a dry soil. Rubber-tyred rollers are most suited to compacting uniformly graded sands. When used to compact cohesive soils, rubber-tyred rollers give the best performance when the soil is about 2–4% below the plastic limit.

Maximum dry density can be obtained with rubber-tyred rollers by compacting layers of soil, 125–300 mm in thickness, with from 4 to 8 passes. These rollers generally can compact soil in less time and at lower cost than sheepsfoot rollers.

Compaction by a rubber-tyred roller is sensitive to the moisture content of the soil. For instance, if the moisture content is on the wet side of the laboratory determined optimum, then this necessitates an increase in the number of passes of a rubber-tyred roller in order to provide a given soil density. The optimum moisture content required for compaction of a soil by a rubber-tyred roller occurs at a higher degree of saturation than that for a sheepsfoot roller. This may be detrimental in

Table 6.1 Typical compaction characteristics for soils used in earthwork construction (from BS6031:1981). The information in this table should be taken only as a general guide. When the material performance cannot be predicted, it may be established by earthwork trials. This table is applicable only to fill placed and compacted in layers. It is not applicable to deep compaction of materials *in situ*.

Soil	Major divisions	Subgroups	Suitable type of compaction plant	Minimum number of passes for satisfactory compaction	Maximum thickness of compacted layer (mm)	Remarks
Coarse soils	Gravel sand gravelly soils	Well-graded gravel and gravel/sand mixtures; little or no fines Well-graded gravel/sand mixtures with excellent clay binder Uniform gravel; little or no fines Poorly graded gravel and gravel/sand mixtures; little or no fines Gravel with excess fines, silty gravel, clayey gravel, poorly graded gravel/sand/clay mixtures	Grid roller over 540 kg per 100 mm of roll Pneumatic-tyred roller over 2000 kg per wheel Vibratory plate compactor over 1100 kg/m² of baseplate Smooth-wheeled roller Vibratory roller Vibro-rammer Self-propelled tamping roller	3–12 depending on type of plant	75–275 depending on type of plant	
	Sands and sandy soils	Well-graded sands and gravelly sands; little or no fines Well-graded sands with excellent clay binder				
	Uniform sands and gravels	Uniform gravels; little or no fines Uniform sands; little or no fines Poorly graded sands; little or no fines Sands with fines, silty sands, clayey sands poorly graded sand/clay mixtures	Smooth-wheeled roller below 500 kg per 100 mm of roll Grit roller below 540 kg per 100 mm Pneumatic-tyred roller below 1500 kg per wheel Vibratory roller Vibrating plate compactor Vibro-tamper	3–16 depending on type of plant	75–300 depending on type of plant	
Fine soils	Soils having low plasticity	Silts (inorganic) and very fine sands, silty or clayey fine sands with slight plasticity Clayey silts (inorganic) Organic silts of low plasticity	Sheepsfoot roller Smooth-wheeled roller Pneumatic-tyred roller Vibratory roller over 70 kg per 100 mm of roll Vibratory plate compactor over 1400 kg/m² of baseplate Vibro-tamper Power rammer	4–8 depending on type of plant	100–450 depending on type of plant	If moisture content is low it may be preferable to use a vibratory roller Sheepsfoot rollers are best suited to soils at a moisture content below their plastic limit
	Soils having medium plasticity	Silty and sandy clays (inorganic) of medium plasticity Clays (inorganic) of medium plasticity Organic clays of medium plasticity				Generally unsuitable for earthworks Should only be used when circumstances are favourable
	Soils having high plasticity	Fine sandy and silty soils, plastic silts Clay (inorganic) of high plasticity, fat clays Organic clays of high plasticity				Should not be used for earthworks

Note: If earthworks trials are carried out, the number of field density tests on the compacted material should be related to the variability of the soils and the standard deviation of the results obtained. Compaction of mixed soils should be based on that subgroup requiring most compactive effort.

embankments where high construction pore-water pressures cannot be tolerated. On the other hand, the construction of earth embankments in regions where rainfall frequently occurs is expedited by the use of rubber-tyred rollers since they help seal the surface of the compacted soil and thereby reduce infiltration.

Smooth-wheeled rollers are most suitable for compacting gravels and sands (Table 6.1). In granular soils, the control of moisture content is important in that it should be adjusted to the optimum moisture content. The depth of the layer to be compacted is governed by the nature of the work as well as the weight of the roller. Generally, however, individual layers vary in thickness, from 50 mm for sub-grades to 450 mm for the base of embankments. Smooth-drum rollers are not recommended for compacting cohesive soils in earth dams because of their low unit pressures and the smooth surface they produce.

Vibratory rollers may be equipped with rubber tyres, smooth-wheeled drums or tamping feet (Lewis and Parsons, 1958). Vibration must provide sufficient force (dead weight plus dynamic force) acting through the required distance (amplitude) and give sufficient time for movement of grains (frequency) to take place. The thickness of each compacted layer is governed by the vibration frequency and weight of a vibratory roller. Lightweight, high-frequency rollers obtain satisfactory densities in thin lifts, while heavyweight, low-frequency rollers obtain satisfactory densities in thick lifts.

The speed and number of passes of a vibratory roller are critical, inasmuch as they govern the number of dynamic load applications developed at each point of the compacted fill. Increasing the number of passes increases the compactive effort as well as the effective depth of compaction (Figure 6.2). Nonetheless, after a few passes, a further increase in depth of compaction requires many passes.

Vibratory rollers have been used successfully for compacting sand, gravel and some cohesive soils (Toombes, 1969; Table 6.1). The frequency and deadweight of vibratory rollers must be suited to the material being compacted. For instance, heavyweight rollers with low-frequency vibrations are used to compact gravel; light to medium-weight rollers with high-frequency vibrations are used for sands; and heavyweight rollers with low-frequency vibrations are used for clays. The best compaction is obtained with vibratory rollers when the soil is at or slightly wetter than optimum moisture content.

Rammers can compact relatively thick layers of granular soils. However, when they are used to compact clay, each layer should not exceed 225 mm in thickness. The moisture content of soil compacted by a heavy rammer should be maintained just below the optimum moisture content determined by the standard compaction test. Because rammers have a

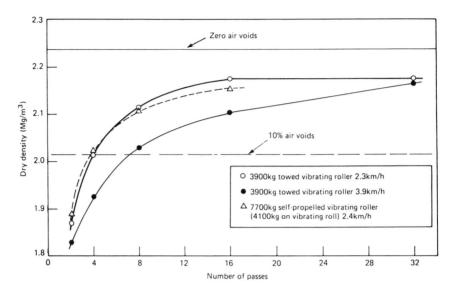

Figure 6.2 Relationship between dry density of 150 mm thick layer and number of passes of vibrating rollers on a well-graded sand with a moisture content of 7.5% (after Lewis and Parsons, 1958; Toombs, 1959) (with permission, the Transport Research Laboratory, Berks)

low output their use is generally restricted to special sites where rollers cannot be employed, such as trenches and behind bridge abutments.

6.1.3 Compaction and soil types

The engineering properties of soils used in fills, such as their shear strength, consolidation characteristics and permeability, are influenced by the amount of compaction they have undergone. Therefore, the desired amount of compaction is established in relation to the engineering properties required for the fill to perform its design function. A specification for compaction needs to indicate the type of compactor, mass, speed of travel, and any other factors influencing performance such as frequency of vibration, thickness of individual layers to be compacted and number of passes of the compactor. Table 6.2 provides a choice of compaction plant for the construction of earthworks. Procedures for the use of compaction plant with different soil types are provided, as are the number of passes and thickness of the layer designed to give a compactive effort capable of achieving an adequate state of compaction with the more difficult soil conditions likely to be encountered.

The compaction characteristics of clay are largely governed by its

Table 6.2 Three methods specified for the compaction of earthwork materials (Anon., 1986a)

Type of compaction plant	Category	Method 1 (wet cohesive soil)		Method 2 (well-graded granular and dry or stony cohesive soils)		Method 3 (uniformly graded granular and silty cohesive soils)	
		D	N	D	N	D	N
Smooth-wheeled roller	Mass per metre width of roll:						
	over 2100 kg up to 2700 kg	125	8	125	10	125	10*
	over 2700 kg up to 5400 kg	125	6	125	8	125	8*
	over 5400 kg	150	4	150	8	Unsuitable	
Grid roller	Mass per metre width of roll:						
	over 2700 kg up to 5400 kg	150	10	Unsuitable		150	10
	over 5400 kg up to 8000 kg	150	8	125	12	Unsuitable	
	over 8000 kg	150	4	150	12	Unsuitable	
Tamping roller	Mass per metre width of roll:						
	over 4000 kg	225	4	150	12	250	4
Pneumatic-tyred roller	Mass per wheel:						
	over 1000 kg up to 1500 kg	125	6	Unsuitable		150	10*
	over 1500 kg up to 2000 kg	150	5	Unsuitable		Unsuitable	
	over 2000 kg up to 2500 kg	175	4	125	12	Unsuitable	
	over 2500 kg up to 4000 kg	225	4	125	10	Unsuitable	
	over 4000 kg up to 6000 kg	300	4	125	10	Unsuitable	
	over 6000 kg up to 8000 kg	350	4	150	8	Unsuitable	
	over 8000 kg up to 12 000 kg	400	4	150	8	Unsuitable	
	over 12 000 kg	450	4	175	8	Unsuitable	
Vibrating roller	Mass per metre width of a vibrating roll:						
	over 270 kg up to 450 kg	Unsuitable		75	16	150	16
	over 450 kg up to 700 kg	Unsuitable		75	12	150	12
	over 700 kg up to 1300 kg	100	12	125	12	150	6
	over 1300 kg up to 1800 kg	125	8	150	8	200	10*
	over 1800 kg up to 2300 kg	150	4	150	4	225	12*
	over 2300 kg up to 2900 kg	175	4	175	4	250	10*
	over 2900 kg up to 3600 kg	200	4	200	4	275	8*
	over 3600 kg up to 4300 kg	225	4	225	4	300	8*
	over 4300 kg up to 5000 kg	250	4	250	4	300	6*
	over 5000 kg	275	4	275	4	300	4*
Vibrating-plate compactor	Mass per unit area of baseplate:						
	over 880 kg up to 1100 kg	Unsuitable		Unsuitable		75	6
	over 1100 kg up to 1200 kg	Unsuitable		75	10	100	6
	over 1200 kg up to 1400 kg	Unsuitable		75	6	150	6
	over 1400 kg up to 1800 kg	100	6	125	6	150	4
	over 1800 kg up to 2100 kg	150	6	150	5	200	4
	over 2100 kg	200	6	200	5	250	4
Vibro-tamper	Mass:						
	over 50 kg up to 65 kg	100	3	100	3	150	3
	over 65 kg up to 75 kg	125	3	125	3	200	3
	over 75 kg	200	3	150	3	225	3
Power rammer	Mass:						
	100 kg up to 500 kg	150	4	150	6	Unsuitable	
	over 500 kg	275	8	275	12	Unsuitable	
Dropping-weight compactor	Mass of rammer over 500 kg: Height of drop:						
	over 1 m up to 2 m	600	4	600	8	450	8
	over 2 m	600	2	600	4	Unsuitable	

* Rollers should be towed by track-laying tractors. Self-propelled rollers are unsuitable.
D = maximum depth at compacted layer (mm). N = minimum number of passes.

moisture content (Hilf, 1975). For instance, a greater compactive effort is necessary as the moisture content is lowered. It may be necessary to use thinner layers and more passes by heavier compaction plant than required for granular materials. The properties of cohesive fills also depend to a much greater extent on the placement conditions than do those of a coarse-grained fill.

The shear strength of a given compacted cohesive soil depends on the density and the moisture content at the time of shear. The pore-water pressures developed while the soil is being subjected to shear are also of great importance in determining the strength of such soils. For example, if a cohesive soil is significantly drier than optimum moisture content, then it has a high strength owing to high negative pore-water pressures developed as a consequence of capillary action. The strength declines as the optimum moisture content is approached and continues to decrease on the wet side of optimum. Increases in pore-water pressures produced by volume changes coincident with shearing tend to lower the strength of a compacted cohesive soil.

Furthermore, there is a rapid increase in pore-water pressures as the moisture content approaches optimum. Nevertheless, compaction of cohesive soil with moisture contents that are slightly less than optimum frequently gives rise to an increase in strength, because the increase in the value of friction more than compensates for the change in pore-water pressure.

The compressibility of a compacted cohesive soil also depends on its density and moisture content at the time of loading. However, its placement moisture content tends to affect compressibility more than does its dry density. If a soil is compacted significantly dry of optimum, and is then saturated, extra settlement occurs on loading. This does not occur when soils are compacted at optimum moisture content or on the wet-side of optimum.

A sample of cohesive soil compacted on the dry-side of optimum moisture content swells more, at the same confining pressure, when given access to moisture than a sample compacted on the wet-side. This is because the former type has a greater moisture deficiency and a lower degree of saturation than the latter, as well as a more random arrangement of particles. Even at the same degrees of saturation, a soil compacted on the dry-side tends to swell more than one compacted wet of optimum. Conversely, a sample compacted wet of optimum shrinks more on drying than a soil, at the same density, that has been compacted dry of optimum.

When expansive clays must be used as compaction materials they should be compacted as wet as is practicable, consistent with compressibility requirements. Certainly it would appear that compaction wet of optimum moisture content gives rise to lower amounts of swelling and swelling pressure. But this produces fill with lower strength and higher compressibility. Therefore, the choice of moisture content and dry density should not be based solely on lower swellability. Consideration should be given to adding various salts to reduce the swelling potential by increasing the ion concentration in the pore water.

A minimum permeability occurs in a compacted cohesive soil that is at or is slightly above optimum moisture content, after which a slight increase in permeability occurs. The noteworthy reduction in permeability that occurs with an increase in moulding water on the dry-side of optimum is brought about by an improvement in the orientation of soil particles, which probably increases the tortuosity of flow, and by the decrease in the size of the largest flow channels. Conversely, on the wet-side of optimum moisture content, the permeability increases slightly since the effects of decreasing dry density more than offset the effects of improved particle orientation. Increasing the compactive effect lowers the permeability because it increases both the orientation of particles and the compacted density.

The moisture content has a great influence on both the strength and compaction characteristics of silty soils. For example, an increase in moisture content of 1 or 2%, together with the disturbance due to spreading and compaction, can give rise to very considerable reductions in shear strength, making the material impossible to compact.

Because granular soils are relatively permeable, even when compacted, they are not affected significantly by their moisture content during the compaction process. In other words, for a given compactive effort, the dry density obtained is high when the soil is dry and high when the soil is saturated, with somewhat lower densities occurring when the soil has intermediate amounts of water (Hilf, 1975). Moisture can be forced from the pores of granular soils by compaction equipment and so a high standard of compaction can be obtained even if the material initially has a high moisture content. Normally granular soils are easy to compact. However, if granular soils are uniformly graded, then a high degree of compaction near to the surface of the fill may prove difficult to obtain, particularly when vibrating rollers are used. This problem usually is resolved when the succeeding layer is compacted in that the loose surface of the lower layer is also compacted. According to Anon. (1981(a)) improved compaction of uniformly graded granular material can be brought about by maintaining as high a moisture content as possible by intensive watering and by making the final passes at a higher speed using a non-vibratory smooth-wheeled roller or grid roller. When compacted granular soils have a high load-bearing capacity and are not compressible they are not usually susceptible to frost action unless they contain a high proportion of fines. Unfortunately, however, if granular material contains a significant

Compaction techniques

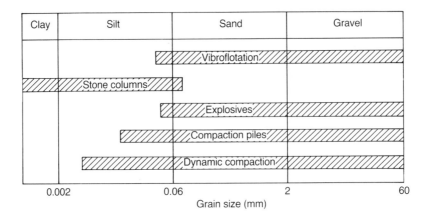

Figure 6.3 Applicability of deep compaction methods (after Mitchell, 1970, with permission of ASCE)

amount of fines, then high pore-water pressures can develop during compaction if the moisture content of the soil is high.

It is important to provide an adequate relative density in granular soil that may become saturated and subjected to static or, more particularly, to dynamic shear stresses. For example, a quick condition might develop in granular soils with a relative density of less than 50% during ground accelerations of approximately 0.1 g. On the other hand, if the relative density is greater than 75%, liquefaction is unlikely to occur for most earthquake loadings. Consequently, it has been suggested that in order to reduce the risk of liquefaction, granular soils should be densified to a minimum relative density of 85% in the foundation area and to at least 70% within the zone of influence of the foundation.

Whetton and Weaver (1991) demonstrated that intensive surface compaction by a heavy vibratory roller (10 750 kg), operating at 30 Hz, can be used to compact granular fill, the required density to avoid liquefaction being achieved by 10–12 passes at a forward speed of 0.3–0.6 m/s. The layer compacted was 1.8 m thick of saturated gravelly sand. The greatest density within the layer was obtained within the depth interval 0.6–1.2 m, relative density exceeding 85%. The lesser densification above was explained as due to insufficient overburden pressure for the selected vibration level and compactive energy. This produces a loosening effect due to larger vibration of particles which has been referred to as overcompaction or overvibration. Whetton and Weaver maintained that the maximum depth of improvement increases with roller energy but that the location of the water table, fines content or presence of any hard material that reflects vibrations can have a significant effect on the maximum depth and effectiveness of densification (the former increases while the latter two decrease the depth of densification). Nonetheless, the vibration energy from intensive surface compaction attenuates fairly rapidly with depth and the degree of densification drops accordingly. The compaction of layers of clean sand up to 3 m in thickness by heavy vibratory rollers had previously been described by D'Appolonia *et al.* (1969) and Moorhouse and Baker (1969).

Thick deposits of soft soils or loosely packed cohesionless soils require improvement in order to reduce the amount of settlement, which occurs on loading, to acceptable amounts. Methods used to bring about *in-situ* compaction of soils include precompression, compaction piles, vibro-compaction, dynamic compaction and explosive compaction. Their applicability to various soil types is shown in Figure 6.3.

6.2 PRECOMPRESSION

Precompression involves compressing the soil under an applied pressure prior to placing a load (Aldrich, 1965). As such, it has proved an effective means of enhancing the support afforded shallow foundations and commonly is used for controlling the magnitude of post-construction settlement. Precompression is well suited for use with soils which undergo large decreases in volume and increases in strength under sustained static loads when there is insufficient time for the required compression to occur. Those soils that are best suited to improvement by precompression include compressible silts, saturated soft clays, organic clays and peats.

Precompression is normally brought about by preloading, which involves the placement and removal of a dead load (Johnson, 1970). This compresses the foundation soils thereby inducing settlement prior to construction. If the load intensity from the dead weight exceeds the pressure imposed by the final load, then this is referred to as *surcharging*. In other words, the surcharge is the excess load additional to the final load and is used to accelerate the process of settlement. The ratio of the surcharge load to the final load is termed the *surcharge ratio*. The surcharge load is removed after a certain amount of settlement has taken place. The installation of vertical drains beneath the precompression load helps shorten the time required to bring about primary consolidation. Earthfill, and to a lesser extent rockfill, is the most frequently used material for precompression.

As stated, the objective of preloading and the length of time it is applied is to reduce the amount of settlement that occurs after the construction period has ended (Figure 6.4). To this end the amount of preloading needed to give the required settlement at final pressure in a certain time is ascertained or, alternatively, the time required to produce a given amount of settlement under a particular preload is determined.

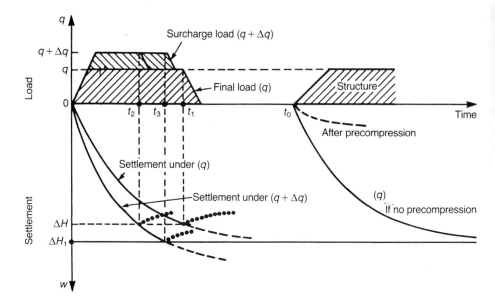

Figure 6.4 Principle of precompression using surcharge loading (i.e. load in excess of final load to accelerate settlement). The length of time to achieve the required amount of settlement has to be assessed, or alternatively the magnitude of the surcharge, $\triangle q$, required to ensure that settlement, $\triangle H$, will be completed in time, t_2, has to be determined. If the surcharge is left in place until time t_2, then it will give the same amount of settlement as that which would occur under the final load at time t_1.

The rate of time of settlement for one-dimensional primary consolidation can be obtained by using the method given by Terzaghi (1943). However, there are certain situations in which the Terzaghi theory does not apply, as, for example, where the strain profile is not constant with depth (Mitchell, 1981). Where surcharging is involved in precompression the degree of consolidation, $U_{q+\Delta q}$, developed under final loading, q, and surcharge, Δ_q, can be derived from

$$U_{q+\Delta q} = \frac{\Delta H}{\Delta H_{q+\Delta q}}$$

(6.1)

where ΔH and $\Delta H_{q+\Delta q}$ are derived from

$$\Delta H_1 = \frac{H}{1 + e_0} \, C_c \log \frac{\sigma'_0 + q}{\sigma'_0}$$

(6.2)

and

$$\Delta H_{q+\Delta q} = \frac{H}{1 + e_0} \, C_c \log \frac{\sigma'_0 + q + \Delta_q}{\sigma'_0}$$

(6.3)

respectively, and in which

H = thickness of the layer
e_0 = initial void ratio
C_c = compressibility index
σ'_0 = initial effective overburden pressure.

When the necessary degree of consolidation, U, has been determined for assumed values of permanent loading, the corresponding time factor, T_v, can be determined from Figure 6.5(a). The time, t, for surcharge removal is then derived from:

$$t = \frac{T_v H^2}{c_v} \tag{6.4}$$

where c_v is the coefficient of consolidation. If rebound following surcharge removal is to be kept to a minimum, then the amount of permanent load preferably should not be less than approximately one-third that of the surcharge.

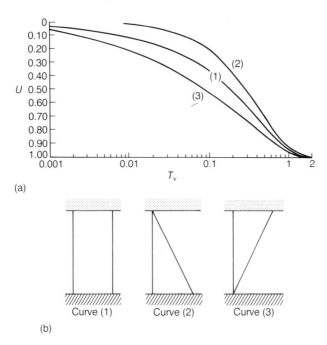

Figure 6.5 (a) Time factor, T_v, versus degree of consolidation, U. Where the consolidating layer drains from both the top and bottom, that is, the layer is open, then curve (1) is used. Where drainage takes place in one direction only, that is, the consolidating layer is half-closed, the curve used depends upon the initial excess pore-water pressure as shown in (b). (b) Initial variations in excess pore water pressure in a half-closed layer. If the layer is half-closed, then $H/2$ is used in eq. (6.4).

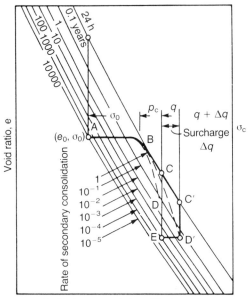

Vertical pressure in logarithmic scale

Figure 6.6 Surcharge to reduce rate of secondary compression (after Bjerrum, 1972, with permission of ASCE). It is assumed that $\sigma'_0 + q + \Delta q$ is well above pre-compression pressure, p_c, in each layer (where σ'_0 is the initial effective stress) so that the representative point C in the e-log p graph is on the virgin curve. Because secondary compression may cause large amounts of total settlement in some soils (mostly organic clays and peat), the effect of surcharge must be analysed with reference to this secondary compression. If one considers the e-log p graph, the primary compression path is ABC and secondary compression can be represented according to Bjerrum by CD. In most cases the actual path is represented by the dotted line between B and D because secondary compression starts during the primary compression period. A precompression project with surcharge is represented by $ABC'D'$ (or ABD') and surcharge removal is $D'E$. The degree of secondary consolidation can be read on the set of e-log p curves giving the equilibrium void ratio at different times of sustained loading (0.1 year, 1 year, 10 years, etc.). Hence, due to surcharge, a much greater degree of secondary consolidation (100 years) can be achieved than without: a state which is represented by point D (6 months)

Secondary compression may cause significant amounts of settlement in some soils (mainly organic clays and peat). Hence the effect of surcharge must also take account of secondary compression (Figure 6.6). The rate of secondary compression appears to decrease with time in a logarithmic manner and its amount is directly proportional to the thickness of a compressible layer at the start of secondary compression.

Johnson (1970) noted that secondary compression appears to be due to shear stresses and that settlements therefore beneath the edges of fills where the shear stresses are high may exceed values beneath the central part of preload fills. An approximate procedure for reducing the amount of secondary compression that occurs after construction involves adding a secondary compression time settlement curve to the primary compression curve in which the amount of secondary compression, ΔH_2, is derived from

$$\Delta H_2 = C_2 H \log \frac{t_s}{t_c} \qquad (6.5)$$

where

C_s = the coefficient of secondary compression (it can be regarded as the ratio of decrease in sample height to the initial sample height for one cycle of time on logarithmic scale following completion of primary consolidation)

H = initial thickness of compressible stratum

t_s = the useful life of the structure or time for which the settlement is significant

t_c = time for the completion of primary consolidation.

In the majority of cases where precompression has been used, it has involved soft fine-grained soils that were either normally consolidated or only slightly overconsolidated. Precompression, as remarked, also has been used to treat peaty soils (Sasaki, 1985). The time available for compression generally limits the practice to relatively thin layers and to soil types that compress rapidly. Soil undergoes considerably more compression during the first phase of loading than during any subsequent reloading. Moreover, the amount of expansion or heave following unloading is not significant.

The geological history of a site and details regarding the types of subsoil, its stratification, strength and compressibility characteristics are of greater importance as regards the successful application of precompression techniques than they are for alternative methods. As can be inferred from above, of particular importance is the determination of the amount and rate of consolidation of the soil mass concerned. Hence sufficient samples must be recovered to locate even thin layers of silt or sand in clay deposits in order to determine their continuity. For example, the presence of thin layers of sand or silt in compressible material may mean that rapid consolidation takes place. Unfortunately, this may be accompanied by the development of abnormally high pore-water pressures in these layers beyond the edge of the fill. This lowers their shearing resistance ultimately to less than that of the surrounding weak soil. Accordingly excess

Compaction techniques

pore-water pressures may have to be relieved by vertical drains.

If a preload fill extends over a long wide area, the influence of layers of sand or silt may be of no consequence as far as accelerating the rate of consolidation beneath the central area of the fill is concerned (Figure 6.7). There consolidation of the soil may be attributable almost exclusively to drainage in a vertical direction. In addition, high pore-water pressures may develop beyond the edges of the fill. As a result, vertical drains or relief wells may be required beneath the central part of large preload fills and beneath their edges.

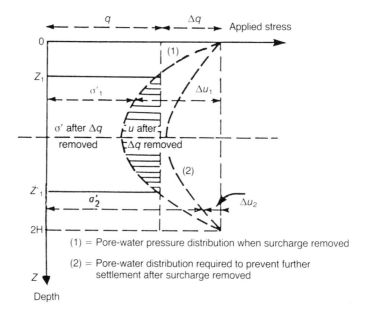

Figure 6.7 Excess pore-water pressure distribution with depth at times of degree of consolidation U_1 and U_2. For a degree of consolidation $U_{q+\Delta q} = U_1$, although ΔH_1 is achieved, there can still be values of excess pore-water pressure, Δu_1, much larger than Δq, in the central part of the layer. On the other hand, in the upper portion (and in the lower portion when both boundaries are pervious) excess pore pressures are much less than Δq. Consequently, after removal of surcharge the upper and lower parts of the clay layer may heave and the central part continues to undergo progressive settlement due to primary consolidation. To eliminate further primary consolidation following the removal of surcharge it is necessary to wait until the degree of consolidation is equal to U_2. Surcharge helps develop secondary consolidation settlement and the degree of secondary consolidation is much higher than for simple precompression. Also, the rate of further consolidation is much smaller. This procedure also assumes that excess pore-water pressures are dissipated in due time and that effective stresses are able to induce proper secondary consolidation

The installation of drains beneath a preload accelerates the rate of settlement in the soil. For example, Robinson and Eivemark (1985) described the use of wick drains and preloading on soft clayey silt. The required settlements under the preload fill were achieved one month after the construction of the preload. Without wick drains a similar amount of settlement would have required up to two years of preloading. Accelerated settlement brought about by sand drains beneath a surcharged fill has also been reported by Moh and Woo (1987).

Precompression also can be brought about by vacuum preloading by pumping from beneath an airtight impervious membrane placed over the ground surface and sealed along its edges. In order to ensure the distribution of the low pressure, a sand layer is placed on the ground beforehand. The 'negative' pressure, created by the pumps, causes the water in the pores of the soil to move towards the surface because of the hydraulic gradient set up. The degree of vacuum that can be obtained depends on the pump capacity and airtightness of the seals (i.e. plastic sheeting, the edges of which are buried in narrow trenches sealed with a bentonite–cement mixture). Values of 60–70% vacuum are generally attained. The consolidation process is similar to that obtained by preloading. It should, however, be emphasized that the low pressure in the voids of the soil increases the size of air bubbles which, in turn, may reduce the permeability of the soil. As in preloading, the method can be improved by use of vertical drains (Woo *et al.*, 1989; Figure 6.8). It can also be used with surcharge loading, the vacuum first being applied to enhance the

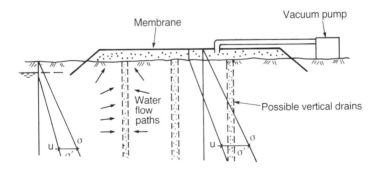

Figure 6.8 Preloading by vacuum. In pressure diagrams σ' is effective pressure and *u* is pore-water pressure. The pressure diagram on the left is for ground conditions prior to treatment; that on the right indicates the changes that occur when vacuum is applied, notably the development of negative pore-water pressure and the increase in effective pressure which lead to consolidation

behaviour of the soil with the surcharge being placed subsequently (Tang and Gao, 1989).

The vacuum method is especially suited to very soft soils where a surcharge may cause instability. Choa (1989) described the use of the vacuum method and band drains to improve very soft to soft silty clay. A possible advantage of the vacuum method is the use of cyclic preloading of the ground, which can lead to an increased rate of settlement. The vacuum method can produce surcharge loads up to 80 kN/m^2.

Water (preponding) has also been used for preloading. Sometimes the ponds may be lined. Water loading in storage tanks is often economical where tanks are to be water tested and where the products that are subsequently to be stored weigh less than water. For example, oil tank farms frequently are located on marshy sites, and in such cases the total tank loads often impose stresses on the soft foundation soils up to ten times their undrained shear strength. Accordingly, controlled water tests have been used to preload foundation soils prior to bringing the oil tanks into service.

Inundation of loess soils has been used in Russia in order to bring about collapse of those types with a metastable soil structure (Evstatiev, 1988). This has included localized shallow wetting of the ground over a number of months from pipelines or shallow excavations where a constant water level is maintained on the one hand and intense deep localized wetting on the other, brought about, for example, by irrigation. If the flow rate in the latter case is high enough to produce a continuous rise in the water table, then saturation of the collapsible soil may occur within several months and the resulting settlement may be extremely uneven. Compaction can be accelerated by the use of vertical sand drains, especially if water is fed into them under pressure. This is the most cost-effective method of treating great thicknesses of loess. However, it may cause some post-treatment deformation which can delay construction, and fissures may develop in the soil.

Consolidation is brought about by dewatering techniques (see Chapter 5). As the water table is lowered this increases the loading on the soil by 9.81 kN/m^2 for every metre of lowering.

6.3 COMPACTION PILES

Soil can be compacted by driving piles into it. Compaction is achieved by simple displacement. However, when the technique is employed in granular soils, compaction is also brought about by the vibratory effect of driving piles, as well as by displacement. Therefore the amount of compaction brought about by compaction piles not only depends on their spacing and size but also upon the effect of the vibration energy.

This is influenced by the content of fines in a deposit since transmission of vibrations is dissipated by the damping effect of fines. In general, if the fines content exceeds 20%, the improvement will be reduced and piles will need to be more closely spaced. Chung *et al.* (1987) suggested an upper fines context of 35% if vibration was going to have an effect on compaction. Compaction piles can be structural piles (timber or concrete) or sand compaction piles.

While densification of loose sand can be brought about by driving any type of displacement piles, sand piles are generally used when the sole purpose is densification (Mitchell, 1970). Sand piles are installed by driving a hollow steel mandrel with a false bottom to the required depth, filling the mandrel with sand, applying air pressure to the top of the sand column, and withdrawing the mandrel. This leaves a column of sand in the hole (Brons and De Kruijff, 1985). During the installation of sand piles the pore-water pressures are increased. However, due to the relatively close spacing of the sand columns and their high permeability, the dissipation of excess pore-water pressures takes place at high rates. The design of sand compaction piles is normally based on the stability of the sand column, and assumes, firstly, that most of the load increase is carried by the sand column which is supported by the surrounding soil, carrying a small increase in load, and, secondly, that the piles and soil are subjected to equal vertical deformation. The sand piles form columns on which the subsequent structural load is concentrated. A well-compacted layer of soil is usually placed on top of the sand piles to obtain a better load concentration on the piles by arching. When

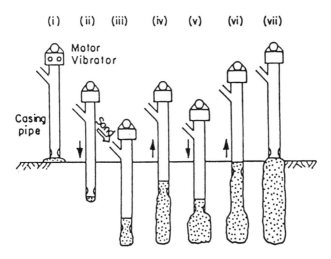

Figure 6.9 Construction of compaction piles by the composer system

Compaction techniques

densifying soils of low permeability below the water table, sand piles also serve as drains.

The vibro-composer method is similar and involves driving a casing to the required depth by means of a vibrator. A quantity of sand is then placed in the pipe which is partially withdrawn as compressed air is blown down the casing (Figure 6.9). Next the pipe is vibrated downwards to compact the sand pile. The process is repeated until the ground surface is reached.

The formation of high-energy input, high-displacement compaction piles involves boring a hole, 0.5 m in diameter, to the required depth in the ground by bailing from within the casing and then ramming sand from it with a drop-hammer, delivering over 100 kJ of energy per blow

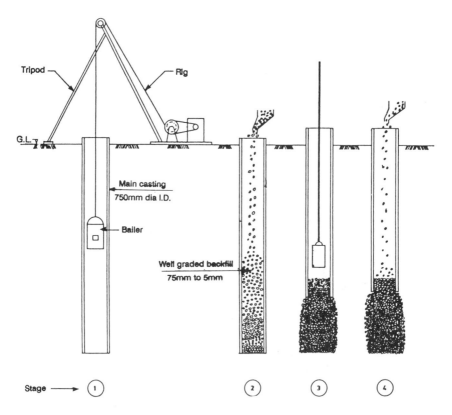

Figure 6.10 Stages in the construction of displacement compaction piles. (1) Boring by bailer with casing to full depth. (2) Pouring granular fill in increments of 2–3 m. (3) Withdrawal of the casing partially and ramming the fill to the specified set; bottom of casing should be at least 0.3 m below the top of the rammed fill. (4) Add further granular fill and repeat stage (3) until the complete length of the column to ground level is formed

(Figure 6.10). The tube is withdrawn in short lifts and at each lift a charge of sand is rammed from the tube. Not only is the sand pile, so formed, highly compacted but it has a considerably greater diameter than that of the tube, thereby compacting the soil around the pile.

The Franki technique of placing gravel compaction piles is similar and includes a tube closed by a plug of gravel (or zero slump concrete) which is driven into the ground by means of a drop-hammer striking the plug. During driving, the soil is compacted radially around the tube. When the tube has been sunk to the design depth an expanded base is formed by ramming a measured quantity of gravel or dry concrete from the tube with a high-energy hammer. The tube is then raised in lifts and further gravel is driven from the tube so forming a continuous pile of compacted gravel. By measuring the quantity of gravel placed in the pile, its diameter may be calculated and, by computing displacement, some measure of the increase in the density of the surrounding soil may be determined. Such piles are constructed on a predetermined spacing that has been established in a test panel before construction begins.

6.4 VIBROCOMPACTION

6.4.1 Vibroflotation in granular soils

Vibrations of appropriate form can reduce intergranular friction among loose cohesionless soils and in this way bring about their densification (Table 6.3). The action allows particles to be rearranged, unconstrained and unstressed, by gravitational forces into the densest possible state. As the grains are rearranged unstressed, no stress–strain readjustment occurs after compaction. Consequently, compaction is permanent. The initial void ratios and compressibilities of granular soils are reduced significantly as a result of vibroflotation and their angles of shearing resistance are increased (Figure 6.11). This, in turn, enhances their bearing capacity, reduces the amount of likely settlement due to loading and increases the resistance of sandy soils to liquefaction. According to D'Appolonia (1953), a relative density of 70% tends to be selected as the criterion for adequate compaction. Compaction not only increases the relative density but it enables the soil to carry heavy or vibrating loads with virtually no settlement since the dynamic stresses involved are less than those produced by the vibroflot (Figure 6.12). It also provides a significant reduction in the risk of settlements during earth tremors.

Vibroflotation is best suited for densifying very loose sands below the water table. The position of the water table is assumed to have no influence on the densities produced. In fact maximum density can be obtained in clean free-draining sands by vibration under either dry (water can be jetted from the vibroflot) or saturated conditions. This is

Table 6.3 Effect of vibroflotation in cohesionless soils (after Mitchell, 1970)

Soil type	Depth (m)	Column spacing (m)	Relative density (%) Before	After
Sand	7	2.0	43	80
Sand and gravel	4.5	–	63	85–95
Well-graded sand	6	2.4	47	79
Sand and gravel	4–5	2.1–2.4	7–58	70–100
Clean loose sand	4	2.3	33	78
Fine sand	5–6	2.0–2.4	0–40	75–93
Gravelly sand	6–9	–	0	80
Sandy gravel	–	–	33–80	85–95
Glacial sand and gravel	5	1.8–2.3	40–60	85–90
Well-graded till	3	3.0	50	75
Loose sand	7	1.9	Loose	80
Loose fine sand	6	1.8–2.3	Loose	80

because a quick condition is developed which reduces the shearing resistance of the sand, thereby facilitating the movement of particles into a denser condition. Liquefaction occurs within the immediate vicinity of the vibroflot, that is, up to a distance of about 500 mm. Beyond this, liquefaction is incomplete because of damping effects.

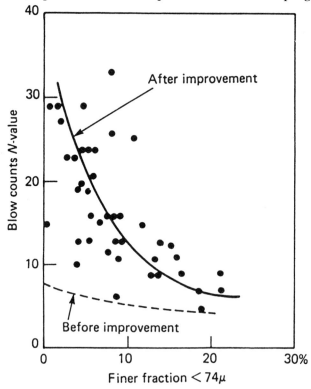

Figure 6.11 Effect of fines content on increase in penetration resistance by vibroflotation (after Saito, 1977)

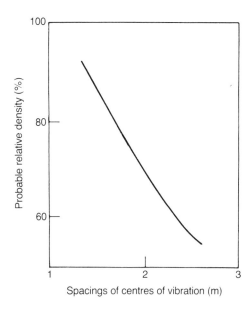

Figure 6.12 Relative density of clean sand at points midway between centres of vibration as a function of probe spacing (after Thorburn, 1975)

On the other hand, the presence of layers of clay, excessive amounts of fines, cementation or organic matter in a soil can cause difficulties for vibroflotation. The densities achieved and the radius of the zone of compaction decrease with increasing silt, clay and organic content (Figure 6.13). Silt, clay and organic material damp the vibrations created by the vibroflot, afford cohesion to particles of sand and fill voids between particles. In this way the relative movement between particles, necessary for densification, is restricted. Layers of clay also reduce the radius of the zone of compaction. Furthermore, clay particles become suspended in the wash water, forming a cake on the walls of the hole and mixing with the backfill material.

The vibroflot, which is usually suspended from a crawler crane (Figure 6.14(a)), can penetrate the ground under its own weight and energy but it is usually assisted by jets of water or air emitted from its base (Figure 6.14(b)). The primary function of jetting, however, is to provide support to the borehole during treatment. Indeed, water is used whenever the borehole formed by the vibroflot is potentially unstable or where groundwater flows into the hole. The water is allowed to overflow from the hole at the surface so that excess hydrostatic pressure and outward seepage forces support the sides.

Compaction of granular soil occurs both during penetration and as the vibroflot is extracted, when it is slowly surged up and down. During

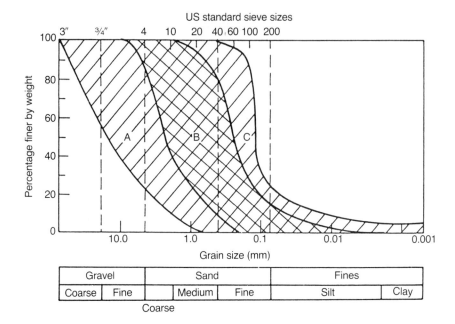

Figure 6.13 Range of particle size distribution suitable for densification by vibro-flotation (after Brown, 1977, with permission of ASCE). (B – most suitable range)

compaction, downward movement of soil immediately surrounding the vibrator creates a cone-shaped depression at the surface which has to be filled continuously with selected granular fill material. The surface may be rolled at the end of treatment. Several interrelated factors influence the densities that can be achieved by vibroflotation (Brown, 1977). These include the type of equipment used, the pattern and spacing of vibroflot centres (Figure 6.15), vibroflot withdrawal procedures, the nature of the backfill and the quality of workmanship. Furthermore, if sands and gravels are to be compacted, they should be sufficiently permeable to permit the release of excess pore-water pressure as they are subjected to vibration. In addition, granular sands should not contain more than 20% fines, of which less than 3% should be active clay. The frequency of the soil vibrations is the same as that of the vibroflot and in order to compact loosely packed granular soils the amplitude and lateral reaction at the end of the vibroflot must be maintained – the larger the amplitude, the larger the radius of influence. The greater the power of vibroflot, the higher is the amplitude of the vibrations. When the area compacted is loaded, the passive resistance of the surrounding soil prevents lateral expansion.

The soil is compacted around the vibroflot to diameters usually

Figure 6.14 (a) Vibroflot suspended from a crawler crane (courtesy of Cementation Ground Engineering Ltd)

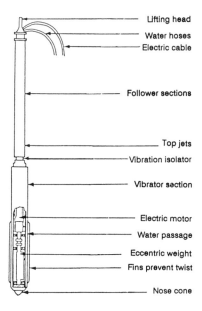

Figure 6.14 (b) Essential features of a vibroflot

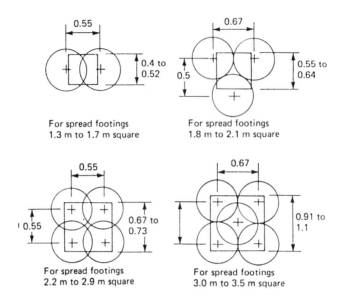

For spread footings
1.3 m to 1.7 m square

For spread footings
1.8 m to 2.1 m square

For spread footings
2.2 m to 2.9 m square

For spread footings
3.0 m to 3.5 m square

Figure 6.15 Typical vibroflotation arrangement for footings (after Brown, 1977, with permission of ASCE)

between 2.4 and 3 m. By inserting the vibroflot in a pattern of overlapping compaction centres, treatment of any required area can be carried out. A triangular grid pattern of treatment centres is usually adopted, the spacing varying according to the site conditions and required bearing capacity. As an example, spacing the centres for vibroflotation 1.8 m apart may develop allowable bearing capacity of 340 kN/m^2 compared with 180 kN/m^2 for a pattern based on 2.7 m centres (Figure 6.16). A square pattern with holes spaced at 1.8–2.25 m centres also may be used for compaction. The fill material, which has to be added to compensate for compaction, increases the dry unit weight of the soil by roughly 10–15%.

Most of the compaction takes place within 2–5 mins at any given depth of the vibroflot in granular soils. It is uneconomic to try to obtain a higher degree of densification by excessively increasing the time of vibration. In order to obtain adequate compaction, it is necessary to supply enough backfill to transmit vibrations from the vibroflot into the surrounding soil and to fill the void left as the vibroflot is withdrawn. Accordingly, the rate of withdrawal of the vibroflot must be consistent with the rate at which backfill material is supplied. If the vibroflot is retracted in small lifts the maximum attainable densities are achieved. For instance, if a vibroflot is raised in increments of 0.3 m and held for

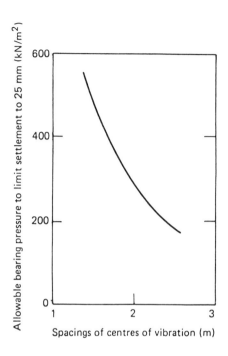

Figure 6.16 Allowable bearing pressure as a function of probe spacing for footings having widths varying from 1 to 3 m (after Thorburn, 1975)

30 s this gives the same withdrawal rate as one raised in 1.2 m increments and held for 2 mins. However, the 1.2 m lift gives more irregular compaction than the 0.3 m lift.

6.4.2 The terra-probe

Granular soils of low density can be compacted by driving and extracting a large open-ended pipe, some 760 mm in diameter, in regular patterns in a deposit using a vibratory pile driver. The pipe is 3–5 m longer than the desired penetration depth. Densification of the soil takes place inside and outside the pipe. The vibrations generated are vertical and the frequency of the vibrator can be varied. Normally, however, the frequency is around 15 Hz. About 15 probes per hour can be carried out at spacings of 1–3 m. The process does not require the addition of backfill around the probe, as occurs in vibroflotation. The increase in density of the soil means that it also undergoes settlement. Consequently, a surcharge of soil usually is added to restore the soil to its original level. It is advantageous to place the necessary surcharge before the compaction operation. The final 0.9–1.5 m is compacted by surface rolling.

6.4.3 Vibroreplacement

Clay soils are largely unaffected by induced vibrations. Nevertheless, vibrocompaction is also used to treat clay soils. In this case stabilization is brought about by columns of densely compacted coarse backfill (Figure 6.17). These stone columns utilize the passive resistance of the soil, thereby increasing its bearing capacity and reducing settlement (Figure 6.18). Stabilization is also aided by the increased rate of dissipation of excess pore-water pressures due to the presence of the columns and the gain in shear strength consequent upon radial drainage (Figure 6.19). Radial drainage means that settlement occurs more rapidly. Stone columns are also a means of reinforcing soil.

A large proportion of any load imposed at the surface is initially borne by the stone columns, the remainder being carried by the soil (Greenwood, 1970; Hughes and Withers, 1974). The vertical displacement of the tops of stone columns within the range of working stresses is less than half the maximum radial strain in the column. Only small radial strains are necessary to develop the minimum passive resistance of the soil since the radial displacement of the soil associated with the formation of the stone columns is considerable. The distribution of the total load between the stone columns and the soil gradually changes until equilibrium conditions are attained. The period of time over which

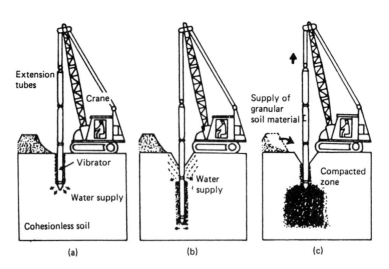

Figure 6.17 The vibrocompaction process (after Baumann and Bauer, 1974, *Canadian Geotechnical Journal* Vol 11.)

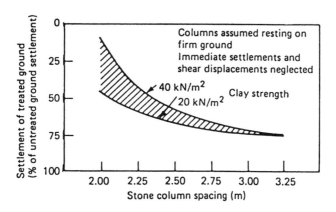

Figure 6.18 The effect of stone columns on anticipated foundation settlement (after Greenwood, 1970)

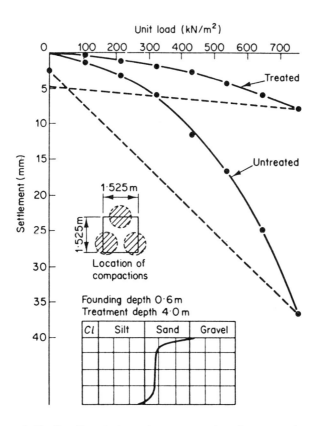

Figure 6.19 Loading test results on treated and untreated sand

this occurs is influenced by the ground and loading conditions and the spacing of the stone columns.

Columns are constructed at numerous centres within clay. The columns are typically 0.9 m in diameter but have varied from 0.5 to 1.5 m. The maximum diameter depends on the properties of the clay. Each column can support a load up to 300 kN.

In soft soils the vibroflot penetrates principally by water jetting in combination with displacement which erodes a hole larger than the machine. The suspension of the vibroflot from a crawler crane ensures that the hole produced is vertical. The vibroflot penetrates the soil to the required depth and is then withdrawn. About 1 m of granular fill is then placed in the bottom of the hole. This is compacted, which tends to displace the surrounding soil laterally (Figure 6.17). The procedure is completed in 1 m lifts up the hole so forming a column of very dense hard granular material. The fill material should vary in particle size from about 5 to 100 mm with not more than 15% material finer than 5 mm.

When backfill is placed with water still flowing up the annular space, vibration transmitted through the backfill to the soil may cause local collapse of surrounding clays, especially if they are sensitive. Collapsed material is removed by the flowing water and the diameter of the hole grows as more backfill is placed.

Considerable excess pore-water pressures are developed in saturated very soft cohesive soils during the initial penetration of the vibroflot, and the soil moves back into position behind the tip of the vibroflot during extraction. Furthermore, very soft cohesive soil liquefies as a result of the high excess pore-water pressures and the high stresses induced by the intense vibrations. It is therefore necessary to form the hole by jetting water from the outlets at the base of the vibroflot. This action flushes out the cohesive soil and prevents the development of high excess pore-water pressures during initial penetration.

Vibroreplacement is commonly used in normally consolidated clays, soils containing thin peat layers, saturated silts and alluvial or estuarine soils (Figure 6.20). Stone columns have been formed successfully in soils with undrained cohesive strength as low as 7 kN/m^2, but normally it is not used in soils with undrained shear strengths of less than 14 kN/m^2 because of the low radial support afforded the stone columns. Vibro-replacement is not recommended as a method of treatment for deep deposits of highly organic silts and clays or thick deposits of peat. Thorburn (1975) maintained that vibroreplacement cannot be relied upon to strengthen made ground which contains large pockets of household refuse or masses of organic matter because the position and extent of these materials cannot readily be established. His experience had shown that radial strains undergone by stone columns due to the presence of these highly compressible materials can cause significant

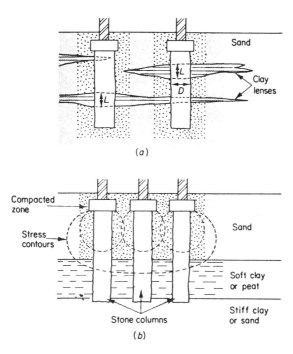

Figure 6.20 Treatment of mixed clays and sands: (a) stone columns stiffening clay lenses; (b) compacted sand raft over soft clay

non-uniform vertical foundation displacements. Made ground containing heavy demolition spoil or timbers can prove difficult or impossible to treat effectively by vibroreplacement.

6.4.4 Concrete columns

Concrete columns, normally 450 mm in diameter and compacted by a vibrator, can be used to support structures on weak cohesive, organic or loose granular soils and fills as an alternative to vibroreplacement. The technique involves the use of a vibrator which contains a special hollow shoe that houses a tremie pipe down which the concrete is pumped. The pipe is connected by hose to a trailer-mounted concrete pump (Figure 6.21). The tremie pipe is initially charged with concrete which is kept under pressure so that the shoe does not become plugged with soil as the ground is penetrated. Vibrating at 3000 Hz, the vibrator is slowly driven to the required depth. In the process weak cohesive soils are improved by lateral displacement and loose sandy soils are compacted. Once at the required depth the vibrator is withdrawn about a metre while pressurized concrete is tremied in at the toe. The vibrator, still

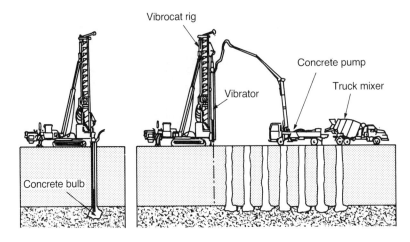

Figure 6.21 Construction sequence of vibrated concrete column

discharging concrete, is then driven back to the original depth to create a bulbous bearing end some 600 mm in diameter. The vibrator, still emitting concrete, is withdrawn at a slower speed than the pumping flow so as to maintain pressure in the concrete column. This forces concrete into the surrounding soil, thus forming a column with a diameter larger than the vibrator. Once grade level is reached the concrete is shut off and the vibrator is briefly re-immersed in the concrete to finish off the column.

6.4.5 Vibrodisplacement

A dry technique of forming stone columns in clays is also widely used. The dry method is employed only where the hole remains stable and where there is no risk of groundwater running into the hole. It is most effective in partially saturated firm clays. The vibroflot penetrates by shearing and displacing the soil around it. Because the vibroflot fits tightly into the borehole there is no annular space into which backfill may be tipped. Consequently the vibroflot has to be removed so that backfill can be placed into the hole. In saturated soft soils the vacuum created as this happens may lead to caving of the sides of the hole. Compressed air is circulated from the tip of the vibroflot to try to prevent this occurring. However, compressed air should not be used if there is standing water in the hole. The combination produces a clay slurry which mitigates against the formation of a sound column.

The shearing action that occurs as soil is displaced as the vibroflot enters the ground in the dry technique tends to break down the structure of the soil around the hole. This may result in the gravel particles in the backfill not being in mutual contact, which is essential for column stability. Shearing displacement also may lead to any free draining layers that are intersected by the column becoming blocked.

Vibrodisplacement accordingly is restricted to strengthening of insensitive soils with an undrained strength of at least 20 kN/m^2 and which, as stated, are unlikely to collapse into the hole.

Displacement is responsible for an increase in the strength of the clay between columns. This does not happen during vibroreplacement. The increase can be up to 1.5 times the original strength for typical column spacing of 1.5 m.

A novel method of using vibrodisplacement to treat peaty soils has been described by Bevan and Johnson (1989). Obviously, conventional vibrodisplacement is not used to treat soils that include significant thicknesses of peat. This is because construction of good continuous columns is difficult through peat but even if these can be formed satisfactorily the surrounding peat offers little lateral support. When load is applied, excessive bulging of the column occurs into the peat. Large and unpredictable settlements then follow. In order to avoid this occurring, trenches are excavated with a minimum of 1 m width and extending 0.5 m below the base of the peat. The trench is then lined with a geogrid which is fixed at the surface. Next granular fill is placed in the trench, the lower 250 mm being 50 mm single size, the rest being filled with sand and gravel. Lastly, the fill is compacted by vibrodisplacement. In this way light or moderately loaded structures may be supported on conventional reinforced strip footings with reinforced ground-bearing floor slabs.

6.5 DYNAMIC COMPACTION

Dynamic compaction is carried out by repeatedly impacting the ground surface by dropping a pounder from a given height from a heavy duty crane at a rate of one blow every 1–3 mins (Figure 6.22). Usually the blows are concentrated at specific locations, the distances between the centres of impact frequently ranging between 4 and 20 m, set out on a grid pattern. The energy per blow is chosen to maximize penetration of the resultant stress impulses (Figure 6.23). Several passes of tamping are required to achieve the desired result. Substantial compaction results, thereby reducing the total and differential settlement that may occur after the erection of structures and permitting the use of spread footings (Gambin, 1987). Dynamic compaction has been used to densify a wide range of soils from organic and silty clay to loosely packed coarse-grained soils and fills (Menard and Broise, 1975). The energy required to achieve a given result increases with the amount of fines in the ratio 1:3 when going from coarse gravel to silty clay.

Figure 6.22 Treatment of suspect ground at Ty Coch, Wales, by dynamic compaction (courtesy of Menard Techniques Ltd)

Dynamic compaction started with weights of about 8 t and drop heights of a maximum of 10 m. Since then the general trend has been to utilize ever-increasing weights and drop heights. At present a machine is in use which has a capacity of lifting 200 t and has a drop height of 40 m (Figure 6.24). It can compact a thickness of 50–60 m of clayey silt. However, dynamic compaction does generate shock waves. Normally the vibration frequency is around 5 Hz, and usually lower than 12 Hz, which represents no danger to existing buildings. Nonetheless, large pounders appear to increase the amplitude of vibrations significantly. The threshold vibration velocity above which damage may occur in buildings is normally taken at 50 mm/s. Even so, damage has occurred at values below this.

An extensive site investigation must be carried out prior to the commencement of dynamic compaction in order that the energy requirements and number of tampings to bring about the necessary improvement in ground conditions can be assessed. A site that is to be treated by dynamic compaction may have to be subdivided into different zones which have different requirements as far as drop height, weight of the tamper and compaction energy are concerned. This depends on the

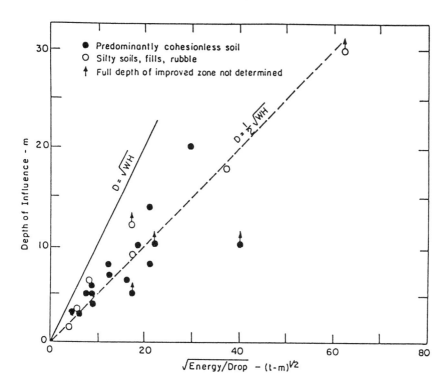

Figure 6.23 Depth of influence as a function of impact energy for dynamic compaction (after Mitchell, 1981)

depth variation of loose soil and on varying requirements placed on the site with regard to settlement. The depth of influence of dynamic compaction depends on impact energy, the damping properties of the soil, the shape of the tamper, the presence of gas in the pores of the soil and dissipation of pore-water pressure.

Dynamic compaction is carried out in several phases (Figure 6.25). The spacing between the points of impact is large in the first pass and is reduced successively in subsequent ones (Hansbo, 1978). The shock waves and high stresses induced by dropping the pounder result in the voids in the soil being compressed, together with partial liquefaction of the soil and the creation of preferential drainage paths through which pore water can be dissipated. This is especially the case in fine sands and silts. The reduction in the void ratio brings about an increase in soil strength and bearing capacity (Figure 6.26). The modulus of deformation of the soil (assessed by the Menard pressuremeter) may be increased by two to ten times which, in turn, means that settlement due to a structural load is reduced.

Figure 6.24 Menard Giga Machine drops 200 tonnes from 22 m in free fall and can compact to a depth of 40 m, Nice airport, 12 m of sand fill resting on alluvium (courtesy of Menard Techniques Ltd)

It is essential to compact the soil at depth first, the soil at the surface being compacted at the end of the operation. It also is important that the depth of the imprint does not exceed half the width of the pounding mass, otherwise an appreciable portion of energy is wasted in kneading the soil and is lost for compaction; hence, it is necessary to have three or four types of tamping weights available on site.

After each pass the craters formed in the ground by tamping are filled with soil (ideally well-graded granular) before the next pass. At the end of the compaction procedure the ground is given a final 'ironing' pass in which the weight is dropped from 1 to 2 m at spacings smaller than the width of the weight.

In practice saturated or partially saturated soils and soft peaty or organic deposits are not subjected to direct dynamic compaction. The standard procedure involves placing a metre or so of fill, such as a sand blanket, over the soil to be treated prior to its compaction. The stress wave produced by the pounder is then somewhat attenuated by the time it reaches the saturated soil. This blanket serves as a bearing layer for the crane and accelerates drainage of any pore water expelled to the surface due to excess pressures. The tamping process both places and mixes the fill into the soft soil beneath, the fill being made up between tamping passes.

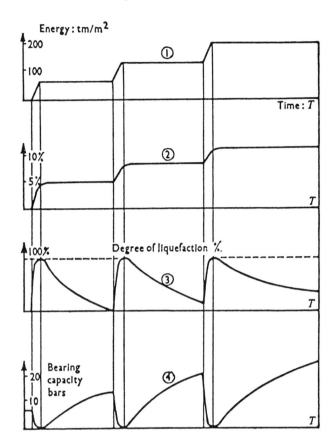

Figure 6.25 Ground response with time after successive coverages of dynamic compaction (after Menard and Broise, 1975). Time between passes varied between one and four weeks, according to soil type. (1) Applied energy in t m/m^2; (2) volume variation with time; (3) ratio of pore-water pressure to initial effective stress; (4) variation of bearing capacity

In the case of saturated soils, silt seems to form the lower safe boundary for compatible materials with regard to grain size. When saturated silty soils are subjected to dynamic compaction, the energy per pass must be limited and several days must be allowed between passes. This permits the dissipation of excess pore-water pressure which develops as the density of the soil is increased. In the case of clayey soil it is recommended that a temporary surcharge should be placed on the soil before it is compacted. If dynamic compaction is carried out in soft clays in association with vertical drains (such as the Geodrain, which can withstand severe dynamic treatment), then this facilitates compaction.

A new, smaller machine has been introduced for ground compaction

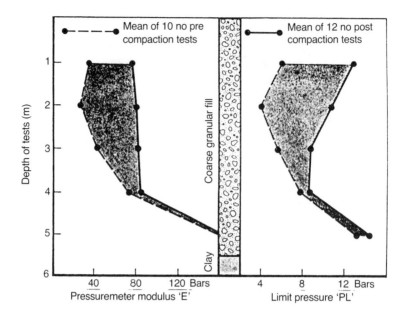

Figure 6.26 Results of pressuremeter tests indicating improvement in strength of ground treated by dynamic compaction

which consists essentially of a large hydraulic pile hammer mounted on a crawler crane (Figure 6.27). Although it can only compact the ground to a depth of some 4 m compared with conventional dynamic compaction methods, which can reach around 20 m, it can be used on small sites

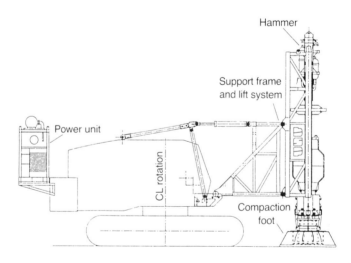

Figure 6.27 Tamping rig attached to crawler crane

near existing buildings, which dynamic compaction cannot. The soil is compacted by a 7 t hammer bearing down on an anvil which is connected by an articulated joint to a foot at its base. The hammer delivers 30 blows per minute. This system ensures that a high percentage of the energy is transferred to the soil. It can be used to improve sites where light structures are to be erected.

6.6 EXPLOSIVE COMPACTION

Explosive compaction tends to be restricted to loosely packed granular soils which contain less than 20% of silt or 5% of clay. For instance, the relative densities of loose sand can be improved by 15–30%. Clay pockets in sands reduce the efficiency of blasting operations. Sudden shocks or vibrations in saturated loose granular soils, cause localized spontaneous liquefaction and rearrangement of the component grains. The load is temporarily transferred to the pore water and the grains are repositioned, adopting a much denser state of packing. This reaction is aided by the weight of the soil above. A saturated condition is very desirable for more uniform propagation of shock waves (Carpentier *et al.*, 1985), otherwise compaction by blasting may be somewhat erratic. Indeed, the pore water acts as a lubricant facilitating the rearrangement of grains. The reduction of the void space in soil results in the displacement of a large volume of water, the quantity depending on the depth, original void ratio and amount of densification attained.

As can be inferred from the above, the most effective results of explosive compaction are obtained in soils that are either dry or completely saturated. As the water content of a soil is reduced from 100% saturation to the absorbed water content around each grain of soil, the increase in capillary tension between the grains means that only a lower state of density can be achieved. This also becomes more pronounced as the soil becomes finer. The effect of capillary tension can be overcome by ponding or flooding the area in question. The presence of only a small amount of gas in the soil can give rise to significant damping of the shock wave.

The distribution of charges, the delay in detonation between them, and their size, should be based upon the required density of the material and the results of field tests. Compaction appears to extend deeper in a layer of soil when small delays are used between charges, that is, only a few seconds. But the amount of compaction is less than with larger intervals. The development of craters at the surface of the soil must be avoided. In addition to the desired amount of densification, the size of the charge is influenced by the type of soil, its thickness, the position of the water table, the dissipation of pore-water pressure, the nearness of structures or open bodies of water, the spacing and overlapping effect of

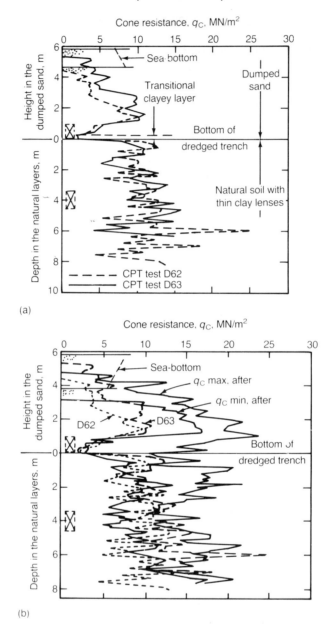

Figure 6.28 (a) CPT tests D62 and D63 in zone B before blasting; (b) minimum and maximum q_c values of tests S10 to S13 after blasting with tests D62 and D63 before blasting (after Carpentier *et al.*, 1985)

charges, and type of explosives used (Pruch, 1963). If compaction is brought about by a series of explosions, then the amount of compaction following each round of blasting is usually less than that of the one preceding.

The maximum depth for effective treatment is limited by the difficulty of placing concentrated charges of sufficient magnitude to generate a shock wave large enough to bring about liquefaction in the soil. Furthermore, as depth increases, so do the effective stresses and strength. Hence the required disruptive stress is greater and the radius of influence decreases. Experience indicates that an explosive charge should be placed at a point approximately two-thirds below the centre of the mass to be densified. It may be necessary to place separate charges in the same blast-hole when granular soils are separated by cohesive soil. In fact thin lenses of clay in granular soils can adversely affect the degree of compaction achieved, reducing it to almost nil (Figure 6.28).

Spacing of blast-holes ranges from 3 to 7.5 m and is governed by the depth of the deposit, the size of the charge and the overlapping effect of adjacent charges. In saturated soils, spacing closer than 3 m should be avoided (Figure 6.29) unless carefully investigated for safety. The charges are placed in holes that are formed by jetting with the aid of casing, and are 50 mm or so in diameter. Then the holes are backfilled.

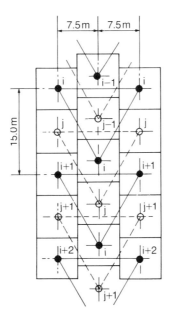

Figure 6.29 Arrangement for blasting. The charges $i, i+1, i+2$ being detonated in the second phase (after Carpentier *et al.*, 1985)

The presence of a surface crust of impervious material inhibits the expulsion of excess pore water from the soil. Hence, it must be removed, disturbed or fractured.

Immediately after the blast the ground surface lifts and is fractured. Gas and water escape from openings, similar to sand boils, which appear at the surface. This may occur over periods ranging from minutes to hours. Primary settlement of the surface follows the initial upheaval more or less immediately and settlement proceeds for some time as the grains of soil continue to be rearranged by the excess pore water that is being expelled. Little densification results in the upper metre or so of soil. This zone, therefore, is compacted by a heavy large mechanical vibrator, where possible.

Blasting has been used to bring about deep compaction in loess soil. Holes are sunk to the required depth and then charged with explosives, the uppermost cartridges being only 500–700 mm below the top of each hole. The charges are detonated separately in each hole, with intervals between blasts of at least 1 min.

The use of preliminary flooding together with deep explosions also has been used to compact loess soil (Litvinov, 1973). Ditches 0.2–0.4 m wide by 4–6 m deep are excavated around the site to isolate the upper

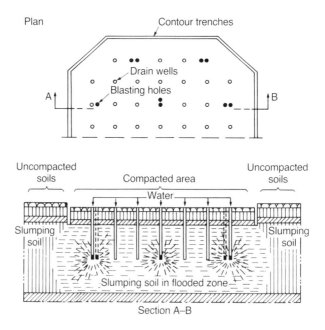

Figure 6.30　Loess compaction by hydro-blasting

zones of the soil in the area to be flooded. Flooding takes several days and may be accomplished by the aid of drains, a few metres apart, sunk into the soil (Figure 6.30). After wetting the soil to a water content exceeding its liquid limit, explosions are set off in boreholes spaced every 3–6 m within the soil. This accelerates compaction since the wet soil is violently shaken which causes the metastable structure to collapse, and settlement of the ground surface then follows. The rate of compaction is some 12 times faster than can be achieved simply by flooding loess, and the degree of compaction obtained is from three to four times greater. Furthermore, the soil is evenly compacted. Reductions in the porosity of soil can be from 50 to 33%.

7

Soil reinforcement and soil anchors

7.1 REINFORCED EARTH

7.1.1 Introduction

Reinforced earth is a composite material consisting of soil containing reinforcing elements which generally comprise strips of galvanized steel or plastic geogrids (Figure 7.1). Soil, especially granular soil, is weak in tension but if strips of material providing reinforcement are placed within it, the tensile forces can be transmitted from the soil to the strips. The composite material then possesses tensile strength in the direction in which the reinforcement runs. The effectiveness of the reinforcement

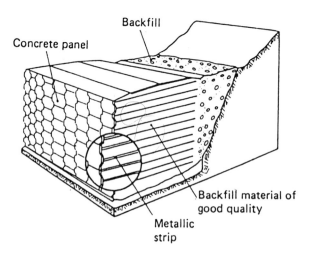

Figure 7.1   Reinforced earth system

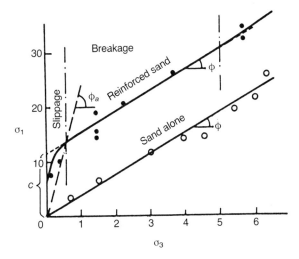

Figure 7.2 Triaxial test on reinforced sand (after Schlosser, 1987). At low values of confining pressure, failure occurs by slippage of the reinforcements. The failure curve can be approximated by a straight line passing through the origin. The strength of the reinforced sand can be represented by an apparent friction angle, ϕ_a, which is greater than ϕ, the internal friction angle of the sand. The apparent friction angle is directly dependent on the density of the reinforcement. The failure curve is a straight line parallel to the failure line of the sand. The strength of the reinforced sand then can be represented by an internal angle of friction, ϕ, and cohesion, c. The value of cohesion is directly proportional to the density of the reinforcement and to it tensile strength

is governed by its tensile strength and the bond it develops with the surrounding soil. Both the shear strength of the soil and the bond strength with the reinforcing elements are frictional and are thus directly related to the normal effective stress distribution. Hence the presence of reinforcement noticeably improves the mechanical properties of granular soil and, depending on the confining pressure, σ_3, two modes of failure can be observed (Figure 7.2). The effectiveness of this composite material depends upon the size, geometry and type of loading of the structure, as well as the types of materials and drainage (Schlosser, 1987).

Reinforcing elements can be made from any material possessing the necessary tensile strength and be of any size and shape which affords the necessary friction surface to prevent slippage and failure by pulling out (Figure 7.3). They should also resist corrosion or deterioration due to other factors. The corrosion/durability of soil-reinforced structures has been discussed in detail by Elias (1990). Steel or aluminium alloy strips, wire mesh, steel cables, glass-fibre-reinforced plastic or polymeric

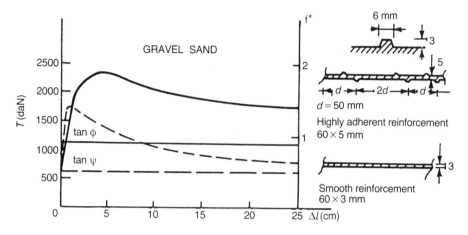

Figure 7.3 Pull-out tests of high adherence and smooth strips (after Schlosser, 1987). The peak and residual values of the apparent coefficient of friction depend on the density of the backfill, the nature of the surface of the reinforcement and the overburden pressure. In dense granular soils, the values of the apparent coefficient of friction, f^*, are usually significantly greater than the values obtained from direct shear tests. This is mainly due to the fact that, under the effect of the shear the soil tends to increase in volume and the denser the fill the more dilatant it will be for a given stress. This volume change is restrained by the surrounding soil and the confining effect results in an increase of the normal stress exerted on the reinforcement and consequently in a high value of the apparent coefficient of friction. The dilatancy effect in a soil can be enhanced by forcing more material to be sheared during the pull-out of the strip. In the case of high adherence (HA) strips this is achieved by the ribs. Another interesting effect of shearing a larger quantity of soil is that the peak, which occurs for a slightly larger displacement, is much flatter. This indicates that the value of the coefficient of friction used in the design of structures is the maximum value for high adherence reinforcements and the residual value for smooth reinforcements. For a given density, a soil will be less dilitant if the confining pressure is higher. This explains why the favourable effect of dilatancy on the apparent coefficient of friction decreases when the average overburden stress (γH) increases. In the case of high adherence reinforcements, the ribs force the shear surface to be within the soil. Thus, under high overburden stress, the coefficient of friction approaches $\tan \phi$, ϕ being the internal friction angle of the soil. In the case of smooth strips, under high over-burden stress, the coefficient of friction approaches $\tan \psi$, ψ being the soil reinforcement friction angle. The value of $\tan \psi$ is approximately equal to $0.5 \tan \phi$. Soil having a soil reinforcement friction angle of less than $22°$ should be rejected

geosynthetic materials (see Chapter 8) have been used as reinforcing elements.

Grid reinforcement systems consist of polymer or metallic elements arranged in rectangular grids, metallic bar-mats, and wire mesh. The two-dimensional grid–soil interaction involves both friction along longitudinal members and passive bearing resistance developed on the cross-members. Hence grids are more resistant to pull-out than strips, however, full passive resistance is only developed with large displacements of the order of 50–100 mm (Schlosser, 1990). The most frequently used polymeric materials in earth-retaining structures are high-density polyethylene and polypropylene grids (see Chapter 8). Several systems of bar and mesh reinforcement have been developed. For example, crude grids have been formed by cross-linking steel-reinforcing bars to form a coarse welded-wire bar-mat. Welded-wire reinforcing mesh similar to that used for reinforcing concrete has also been used to reinforce granular soil, especially in the United States.

Attempts have been made to reinforce soil by including fibre within it. The fibres have included geotextile threads, metallic threads and natural fibres such as bamboo (Mitchell and Villet, 1987). A recent innovation is a three-dimensional reinforcement technique which consists of mixing granular soil backfill with a continuous polymer filament with a diameter of 0.1 mm and a tensile strength of 10 kN (Leflaive, 1988; Schlosser, 1990). Approximately 0.1–0.2% of the composite material consists of filament, resulting in a total length of reinforcement of 200 m per cubic metre of reinforced soil. The technique has been used in soil-retaining walls and has demonstrated both high bearing capacity and resistance to erosion. However, there are difficulties associated with the efficient mixing of fibres with soil. In addition, randomly distributed polymeric mesh elements have been included in soils, the meshes interlocking with the soil particles to strengthen the soil. The ductility and permeability of the soil are not reduced and a relatively homogeneous composite is produced. The ribs of individual mesh elements interlock with groups of soil particles to form an aggregation of partiicles, and adjacent aggregations interlock to form a coherent matrix (Figure 7.4). Numerous types of mesh elements are now being tested in a range of soil types and the practical problems of mixing them with different soils in various situations are being investigated (McGown *et al.*, 1985).

Generally a limited range of soil types have been used as a fill in reinforced earth structures. Indeed, the first specifications referred to clean granular materials. For example, Vidal (1969) maintained that ideally the soil should be composed mainly of granular material with a certain amount of clay. The amount of clay, it was suggested, would vary with the type of reinforcement used, the general condition being that sufficient friction should exist between soil and reinforcement to generate the necessary tensile stresses in the reinforcement. However, the type of fill used is also influenced by the requirements of the reinforced earth structure. As Jones (1985) pointed out, the type of fill

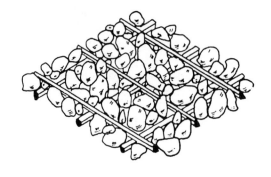

(a)

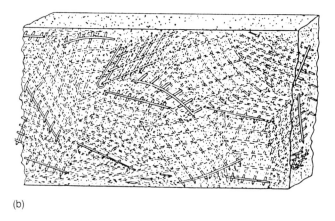

(b)

Figure 7.4 The interlock mechanism for mesh elements (after McGown *et al.*,
1985): (a) interlock with groups of particles; (b) interlock of adjacent aggregations

used for constructing reinforced earth retaining walls needs to be of
better quality than that used for a reinforced earth embankment.

Among the advantages of granular fill are the fact that it is free
draining and non-frost susceptible, as well as being virtually non-
corrosive as far as the reinforcing elements are concerned. It is also
relatively stable, in that its use in reinforced earth more or less elimates
post-construction movements.

It is also necessary to provide some form of barrier to contain the soil
at the edge of a reinforced earth structure (Figure 7.5). This facing can be
either flexible or stiff but it must be strong enough to retain the soil and
to allow the reinforcement to be fixed to it.

Reinforced earth is now widely employed, primarily because the use
of prefabricated elements speeds up the construction process and
because of the lower cost of the technique compared with traditional

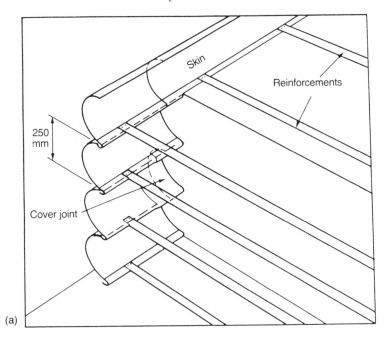

(a)

(b)

Figure 7.5 (a) Diagram showing 'concertina' type facings which were used particularly in earlier reinforced earth. (b) Concrete panels used on a reinforced earth retaining structure on the right-hand abutment of Clyde Dam, New Zealand

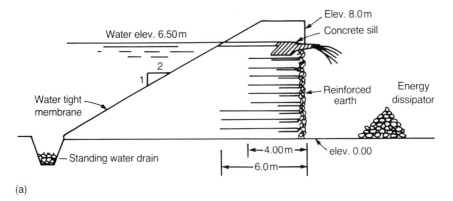

(a)

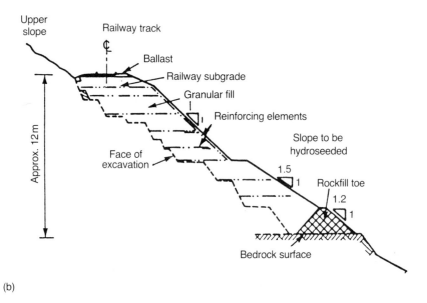

(b)

Figure 7.6 (a) Reinforced earth dam; (b) Stabilization of a slope with reinforced earth

methods. For example, reinforced earth walls cost about half as much as cantilever or crib walls. Yet another reason is the mechanical stability of the material.

Because reinforced earth is flexible and the structural components are built at the same time as backfill is placed, it is particularly suited for use over compressible foundations where differential settlements may occur during or soon after construction. In addition, as a reinforced earth wall uses small prefabricated components and requires no formwork, it can be readily adapted to required variations in height or shape.

Reinforced earth has been used in the construction of embankments, retaining walls and bridge abutments for transportation applications; for industrial structures including material processing and storage facilities, and foundation slabs; for hydraulic structures such as sea walls, flood-protection structures, sedimentation basins and dams (Figure 7.6); and for stabilization and reinstatement of failed slopes (Figure 7.6; Mitchell and Villet, 1987; Mitchell and Christopher, 1990; O'Rourke and Jones, 1990).

7.1.2 Load transfer mechanism and strength development

The load transfer mechanism between the soil and reinforcing element is governed by the limiting friction that is developed at their interface. Because the reinforcing action is derived from friction, it must be determined whether this friction exists without sliding between grains and reinforcement. Andrawes *et al.* (1980) indicated that sliding takes place if there is a tendency for the stress obliquity to increase above that corresponding to the limiting friction–adhesion. Reinforced earth behaves in an elastic manner that is capable of withstanding both internal and external forces.

Because of friction, the upper and lower parts of a layer of soil enclosed between adjacent strips of reinforcement are held by the reinforcement. Consequently, if the strips are placed sufficiently close together, then the complete layer of soil will be laterally restrained and the maximum strain experienced by the soil in the direction of the reinforcements will approximate to the strain in the reinforcing strips. Normally the value of Young's modulus of the reinforcement is much higher than that of the soil, which means that the resulting strains in the soil are so small that it is essentially at rest.

Reinforced earth containing granular fill can be made to resist almost any pressure since the grains usually have a high crushing strength and the reinforcing strips can be placed inside the mass in order to resist external conditions. Furthermore, deformations in the foundation will not give rise to fissures or breaks in such reinforced earth. In such cases the particles slide along the reinforcing strips and thereby achieve another state of equilibrium, generating another stress distribution in the strips. As long as the reinforcing elements can withstand these new loads, the whole structure remains stable. Schlosser and Guilloux (1981) mentioned that many reinforced earth retaining walls built on slopes had been subjected to large deformations and had remained intact. The behaviour of such walls has demonstrated the great flexibility and good performance of reinforced earth materials. Indeed, because reinforced earth can suffer large deformations without rupture it is able to resist external forces such as earthquakes.

Relative extension of the soil compared with the reinforcing strips becomes less for higher walls and their corresponding higher stresses. Hence for higher structures the effective lateral stress is reduced and approaches an active state. Theoretically there is no height limit to reinforced earth structures.

In the case of reinforced earth walls, the wall facing and reinforced earth are considered as a coherent block structure in stability analyses with active earth pressure acting behind the block. The stability of the block is then analysed against sliding, overturning, bearing capacity failure and deep-sealed failure (Manfakh, 1989). The locus of minimum tensile forces in the reinforcements define active and resistant zones within reinforced earth. The effective length of the reinforcements, l_e, is the length to the right of the active core in Figure 7.7. A factor of safety of at least 1.5 is usually required against strip pullout. The minimum tensile stress in the reinforcement must be less than that allowable for the material.

Internal failure can occur only as a result of loss of friction between soil and reinforcement, or by tensile failure of the reinforcing strips. The shear strength of the fill is fully mobilized if the reinforcement breaks. Overstressing could lead to tensile failure of the reinforcement. Determination of sliding shear resistance between soil and reinforcement is not absolutely critical since slippage gives rise to a redistribution of stress and slow deformation of the mass.

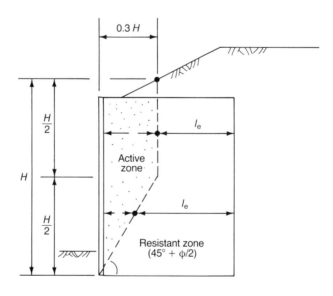

Figure 7.7   Active and resistant zones in a reinforced earth wall

In order to obtain the most benefit from reinforced earth, reinforcing elements should be laid in the direction of the principal strains within the soils. In most reinforced earth structures that have been constructed, the reinforcing elements are laid parallel to each other. Although this has proved a reasonable construction expedient, it means that not all inclusions are aligned in the directions of principal tensile strains or occur in zones of maximum tensile strains for the soil under the same stress condition Thus the effectiveness of the individual layers must vary as their orientations and locations deviate from their optimums. In fact, weakening may occur where the orientation of a layer approaches that of the zero extension lines of the unreinforced soil and the friction between an inclusion and soil is less than that for the soil alone.

7.1.3 Soil types and reinforced earth

Two main factors influence the value of soil-reinforcement friction, namely, the internal angle of friction of the fill material and the roughness of the surfaces of the reinforcing elements. As remarked above, one of the most significant items is the relative volume of the fine-grained fraction of the fill to that of the granular part, since, as the former portion increases, so the internal friction angle decreases. However, as long as the fine-grained fraction is small, the number of grain contacts in the granular skeleton does not vary and so the value of the internal angle of friction remains more or less the same. Schlosser (1987) suggested that, ideally, fill material should not contain more than 15% (by weight) of material smaller than 0.015 mm. He further suggested that the internal friction angle of the saturated and consolidated fill material must be greater than 25°. The assessment of internal friction is not required if no more than 15% (by weight) of the fill material is smaller than 0.08 mm.

Special attention to moisture–density relationships is required when the percentage finer than 0.08 mm is greater than 15%. The compaction specifications should include a specified lift thickness and allowable range of moisture content above and below optimum. In addition, special attention must be given to both internal and external drainage.

The peak and residual shearing resistance developed in reinforcement are dependent on the density of the soil, the effective overburden pressure and the geometry and surface roughness of the reinforcing elements. Higher densities of soil can increase the normal stresses acting on reinforcing elements and the apparent coefficient of friction, at least in those ranges of overburden pressures where the soil is dilatant (i.e. in which the void ratios are less than the critical void ratio). Hence, a knowledge of the void ratio, density, shear strength and state of strain of the soil is important when selecting the apparent friction coefficient. Obviously the nature of the surface of the reinforcement should also be considered.

Granular soils compacted to densities that result in volumetric expansion during shear are ideally suited for use in reinforced earth structures (Vidal, 1969). If these soils are well drained, then effective normal stress transfer between the strips and soil backfill occurs instantaneously as each lift of backfill is placed. Furthermore, the increase in shear strength does not lag behind vertical loading. Granular soils behave as elastic materials within the range of loading normally associated with reinforced earth structures. Therefore, no post-construction movements associated with internal yielding or readjustments, should be anticipated for structures designed at working stress level.

On the other hand, fine-grained materials are not especially suitable for most reinforced earth structures (Vidal, 1969; McKittrick, 1979). The adhesion between fill and reinforcement is poor and may be reduced by an increase in pore-water pressure. Such soils are normally poorly drained and effective stress transfer is not immediate. Hence a much slower construction schedule or an unacceptably low factor of safety in the construction phase is necessary. Moreover, cohesive materials often exhibit elastoplastic or plastic behaviour, thereby increasing the possibility of post-construction movements. Nonetheless, such fill can be used successfully in structures like embankments. Furthermore, Sridharan *et al.* (1991) showed that cohesive soil with low internal friction angles can be used as backfill material when layers of sand are used in contact with the reinforcing strips. The thickness of the layer of sand depends largely on the surface roughness of the reinforcement and the strength of the bulk backfill and to a lesser extent on the shape and size of the reinforcement. Generally, however, a sand layer 15 mm thick is sufficient to increase the interfacial friction angle to one almost equal that of a completely granular backfill. Hence the pull-out resistance, pull-out length behaviour of such a sandwiched system is more or less the same as reinforced earth consisting of granular soil.

7.1.4 Anchored earth

In anchored earth the pull-out resistance is provided by development of bearing pressures around an anchor at the end of the tensile member which connects it to the facing unit (Jones *et al.*, 1985). The ultimate load of an anchor corresponds to the structural failure of the anchor or the anchor pulling through the soil. There also is a difference between the load–deflection characteristics of an anchor and a reinforcing strip. Various forms of anchor can be used, one of the most effective being triangular shaped and formed of mild steel reinforcing bar (Figure 7.8).

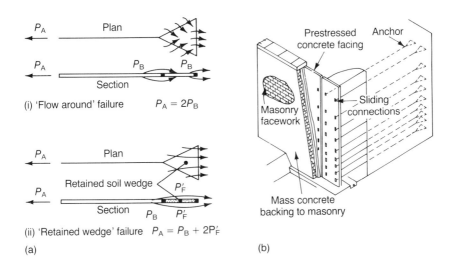

Figure 7.8 (a) The components and shearing mechanisms of a triangular anchor (after Jones *et al.*, 1985). There are two alternative mechanisms for shear around a triangular anchor element. In the first, soil flows around the two leading members and the back member of the anchor as it is pulled through the soil, so that

$$P_A = 2P_B$$

In the second, a triangular wedge of soil is retained within the anchor, shearing occurs on the top and bottom of the triangular wedge and supplements the bearing failure at the two leading members of the anchor, hence

$$P_A = P_B + 2P'_F$$

The value of P_B (appropriate to $\phi' = 35°$), according to Jones *et al.* (1985), may be taken as

$$P_B = 4K_P\sigma_v'Bt$$

where B is the length of the back bar of the anchor which is used to define anchor size, K_P is the coefficient of passive earth pressure, σ_v' is the vertical effective stress and t is the diameter of anchor bar. A lower bound case for

$$P'_F = A\sigma_v' \tan \phi'$$

where A is the plan area of a triangular anchor, ϕ_v' is the vertical effective stress and ϕ' is the peak angle of shearing resistance measured in a shear box

(b) Anchored length of construction using full height facings (after Jones *et al*, 1985)

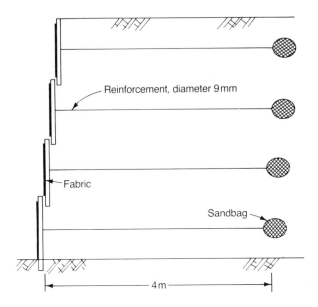

Figure 7.9 Schematic cross-section through reinforced soil wall with multiple tie-bars attached to bags of sand that serve as anchors. The rods and anchors are embedded in the embankment during construction (after Fukuoka, 1986)

An anchor formed from round bar offers the smallest surface area which is advantageous in terms of protecting it against corrosion. The mobilization of bearing pressures around the anchor, rather than dependence on friction along the reinforcement, reduces the need for good frictional fills, thereby allowing lower quality materials to be considered (Temporal *et al.*, 1989). Anchored earth is cheaper than reinforced earth since the quantity of steel used in the anchors is about 60% less than that required for reinforcement.

A similar concept has been developed and used in Japan. For example, Fukuoka and Imamura (1982) described the construction of an anchored earth retaining wall in which each anchor consisted of a 20 mm diameter steel bar attached to a concrete plate, 400 mm x 400 mm, which was embedded in the soil backfill. Subsequently Fukuoka (1986) suggested a similar system in which the tie-bars were attached to sandbags embedded in the backfill during construction (Figure 7.9).

7.2 *IN-SITU* REINFORCEMENT TECHNIQUES

There are three principal types of *in-situ* soil reinforcement techniques used to stabilize slopes and excavations, namely, soil nailing, reticulated micropiling and dowelling. Soil nailing is used to reinforce the ground with small inclusions, usually steel bars. These are installed horizontally or subhorizontally so that they improve the shearing resistance of the soil by acting in tension (Bruce and Jewell, 1986, 1987). The face of the soil mass is stabilized with shotcrete so that the soil nails and shotcrete represent a resistant unit supporting the unreinforced soil behind, in a similar way to a gravity-retaining wall.

Reticulated micropiles are steeply inclined in the soil at various angles both perpendicular and parallel to the face (Lizzi, 1977). Again the aim is similar to that of soil nailing – that is, to provide a stable block of reinforced soil which supports the unreinforced ground. In this method the soil is held together by the multiplicity of reinforcement members acting to resist bending and shearing forces.

Soil dowelling is used to reduce or halt soil creep on slopes. The slopes treated by dowelling are much flatter than those in the other two types of *in-situ* reinforcement.

Reinforcement provides the largest increase in strength when it is angled across the potential shear surface so that the reinforcement is loaded in tension. Hence, where a steep slope is excavated in granular soil the most effective use of reinforcement is to install it through the face in a more or less horizontal direction. Therefore, in such situations soil nailing is likely to be more cost effective than reticulated micropiles. In marginally stable granular or scree slopes, where stability must be improved but where excavation is not foreseen, then either soil nailing or reticulated micropiling is acceptable. As can be inferred from above, in flatter slopes in cohesive soils where stability is governed by a well-defined shear zone, soil dowels are the most appropriate.

7.2.1 Soil nailing

Soil nailing is a technique whereby *in-situ* soil is reinforced by the insertion of steel rods, 20–30 mm in diameter (Gassler and Gudehus, 1981). However, because of the possible corrosion of steel bars, reinforcement coatings with high resistance to corrosion are being developed and fibre-glass nails have been used recently (Gassler, 1992). The technique is employed to enhance the performance of granular soils and stiff clays. Excavations in soft clay (undrained shear strength less than 48 kN/m^2) are unsuited to soil nailing since its low frictional resistance would necessitate a very high density of very long nails to ensure adequate stability. Other soils in which soil nailing is not cost effective include loose granular soils with blowcounts from the standard penetration test of less than 10 or relative densities less than 30%; poorly graded soils with coefficients of uniformity of less than 2 (nailing is not practical because of the necessity of stabilizing the cut face prior to excavation); and highly plastic clays (plasticity index greater than 20%) due to

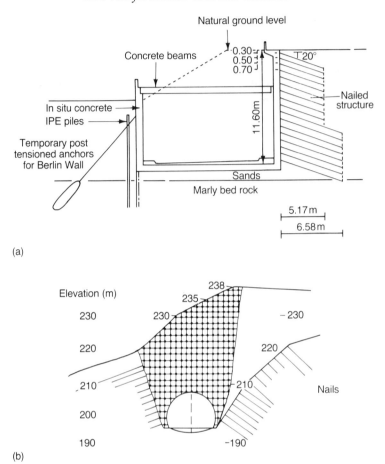

Figure 7.10 (a) Use of soil nailing in the construction of a cut-and-fill tunnel. (b) Use of soil nailing to stabilize the portals of a tunnel and adjacent cut slopes (from Bruce and Jewell, 1987)

excessive creep deformation. It normally is used to improve the stability of slopes and to provide support in excavations (Elias and Juran, 1991; Figure 7.10). As far as nail length is concerned, it may be around 50% of the height of the excavation requiring support. In the case of slope stabilization, nail length depends on the position of the critical shear plane.

Soil nailing is constructed by staged excavations from 'top down' (Figure 7.11). It requires the formation of a series of small cuttings generally 1–2 m high in the soil. Cuts of more than 2 m or less than 0.5 m are rare in granular soils. Greater depths have been used in overconsolidated clays. These cuts must be able to stand unsupported for at

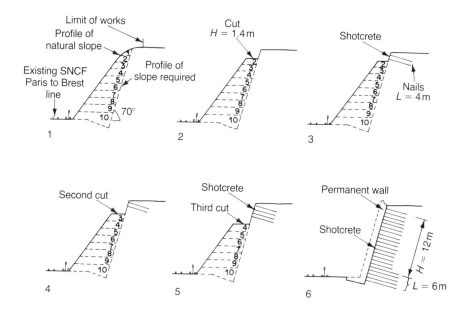

Figure 7.11 Cut slope stabilization construction sequence

least a few hours prior to shotcreting and nailing. The soil must therefore have some degree of cohesion or cementation otherwise a pretreatment such as grouting may be required to stabilize the face. A dewatered face in the excavation is desirable for soil nailing. If groundwater percolates through the face, the unreinforced soil is likely to fail locally on initial excavation, making it impossible to establish a satisfactory coat of shotcrete. As soil nailing is in intimate contact with the excavated soil surface this minimizes the disturbance to the ground. Soil nailing can proceed rapidly and the excavation can be easily shaped. It is a flexible technique that can accommodate variations in soil conditions and work programmes, as excavation progresses.

A level working bench some 6m wide should be available for the operation. Usually the length of a single cut is dictated by the area of the face that can be stabilized during a working shift. Any loosened areas on the face should be removed prior to facing support being applied. Normally the face support must be placed as early as possible to prevent relaxation or ravelling of the soil. The support involves pinning a reinforcement mesh to the face and spraying with shotcrete before drilling the nail holes. The thickness of shotcrete varies from 50 to 150 mm for temporary applications to 150 to 250 mm for permanent projects.

In the 'Hurpinoise' system the angle steel reinforcement is often driven before placing the mesh and shotcrete. A recently introduced system involves inserting the nails into the soil by means of a compressed air 'gun'. Under favourable conditions nails, 38 mm in diameter and up to 6 m in length, can be inserted into the soil at a rate of one every two to three minutes.

A drainage system is required to drain the reinforced zone, and consists of shallow and deep drains. The shallow drains are tubes which convey water from immediately behind the facing. Deep drains consist of slotted tubes and are angled upward at 5–10°. They are usually longer than the nails.

After placement of shotcrete, a series of holes are drilled at, for example, 1.0–1.5 m centres, and are inclined into the soil. The nails are either drilled into the soil or grouted into predrilled holes. Generally borehole diameters range from 75 to 150 mm for drilled and grouted nails. This usually allows a grout annulus of at least 20 mm thickness around the reinforcement providing a degree of corrosion protection. Normally cement grouts are used and are injected at low pressures. In the jet-bolting system very high pressures (over 20 MPa) are used to inject cement grout through small apertures at the top of the nail while it is being installed. This grout lubricates the penetration of the nail and enhances its bond capacity. Alternatively, nails can be driven into the soil. The rate of installation can be very high when nails are driven. However, driven nails may be less suitable in tills and dense cemented soils than grouted nails. Also, care must be taken in loose, weakly cemented granular soils to ensure that driving does not cause local destructuring of the soil around the nail which could result in low values of bond stress. Lastly, the exposed ends of the nails are covered with shotcrete. The whole process is then repeated for the next layer of soil excavated.

A nailed soil structure must be able to resist the outward thrust from the unreinforced soil behind without sliding, and must be stable against deep-seated failure mechanisms. The nails must be installed in a dense enough pattern to ensure an effective interaction with the soil in the reinforced zone, as well as being long enough and strong enough to ensure a stable reinforced zone. Because nails are installed at high density the consequences of one failing is not usually significant. The nail is placed in the ground unstressed and the reinforcement forces are mobilized over its length by subsequent deformation of the soil. As noted above, grouting techniques are usually employed to bond the nail to the surrounding ground. Where the nails are driven, reinforcement forces are sustained by the frictional bond between the soil and the nail. Overall movements needed to mobilize the reinforcement forces are surprisingly small. Because of the flexibility of a soil-nailing system it can resist seismic loading. For instance, three soil-nailing projects in California, which were within 33 km of the epicentre of the Loma Prieta

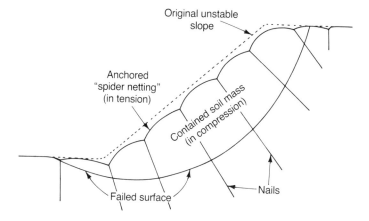

Figure 7.12 Schematic cross-section of anchored geosynthetic 'spider netting' used with soil nails to stabilize a slope (after Koerner and Robins, 1986)

earthquake of 1989, magnitude 7.1, survived undamaged (Ferworn and Weatherby, 1992). Recent accounts of the design of soil nailing have been provided by Bridle (1989), Jewell and Pedley (1990a, 1990b), Bridle and Barr (1990), Juran *et al.* (1990b) and Pedley *et al.* (1990).

A new method of soil nailing uses geotextiles, geogrids or geonets to cover the ground surface (Koerner and Robins, 1986). The geosynthetic material is reinforced at distinct nodes and anchored to the slope using soil nails at the nodes (Figure 7.12). When the nails are properly fastened, they pull the geosynthetic material into the soil, placing it in tension and the constrained soil in compression.

7.2.2 Reticulated micropiles

Small-diameter (75–250 mm) reticulated micropiles, or as they are sometimes called root piles, are installed in cased boreholes. In the smaller diameter range the concrete micropiles are provided with a central reinforcing rod or steel pipe, while those with larger diameters may be provided with a reinforcing bar-cage bound with spiral reinforcement. The principal difference between reticulated micropiles and soil nailing is that, in the former, the reinforcement by the root piles is strongly influenced by their three-dimensional geometric or root-like arrangement (Figure 7.13). The holes are formed by a high-speed rotary drill and, as drilling proceeds, a steel casing is gradually driven into the hole to prevent its collapse and to limit loosening of the soil. Reinforcement, usually a single steel bar, is placed in the centre of the casing. The hole is then filled with a high slump, cement-rich, small aggregate concrete. During withdrawal of the casing additional concrete is placed

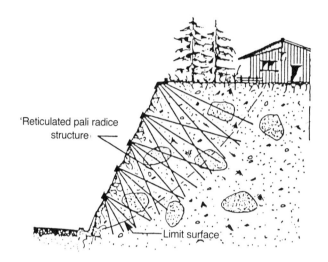

'Reticulated pali radice
structure

Limit surface

Figure 7.13 Reticulated micropiles used for slope stabilization (after Lizzi, 1977)

by pumping or compressed air pressure. This produces an intimate contact between concrete and soil, forces penetration of the concrete into cracks and fissures and, in the case of compressible soils, produces a pile of larger diameter than the original casing. Reticulated micropiles can be installed in gravels, sands, silts and clays. Drilling can take place in restricted areas and does not produce detrimental vibrations.

A cluster of reticulated micropiles, usually containing some installed at a batter, can be used to form a type of *in-situ* reinforced earth mass or monolithic block (Lizzi, 1977). Used in this way these reticulated micropile structures have applications in slope stabilization (Figure 7.13), retaining structures, and underground construction (Figure 7.14).

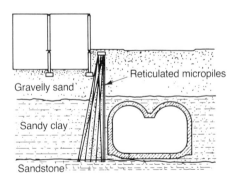

Reticulated micropiles

Gravelly sand

Sandy clay

Sandstone

Figure 7.14 Protection of tunnel by reticulated micropiles

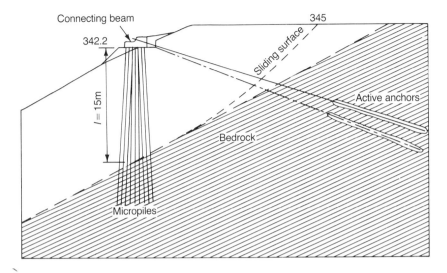

Figure 7.15 Stabilization of a sliding slope with reticulated micropiles and ground anchors attached to connecting beams (after Cantoni *et al.*, 1989)

Individual piles within a reticulated micropile structure may be called on to carry tension, compression and flexural stresses so that the interactions with the included soil are complex. Cantoni *et al.* (1989) described the use of reticulated micropiles, together with ground anchors, to stabilize a sliding slope (Figure 7.15) which was affecting a motorway. Both the micropiles and anchors were attached to connecting beams.

7.2.3 Soil dowels

The stabilization of slopes in stiff clay which are undergoing soil creep by using dowels (Figure 7.16) has been described by Gudehus and Schwarz (1985). They also outlined their design method. The dowels must be distributed so that the soil cannot move between them. The most efficient way to improve the shearing resistance on a weakened shear surface in soil is to use relatively large-diameter dowels (e.g. 1.5 m diameter) with high bending stiffness. Consequently the diameter of a soil dowel is generally far greater than that of a soil nail or micropile. Dowel action only occurs after a lapse of time that is needed for displacement to mobilize a sufficiently high lateral force. The dowels are made of concrete or steel and they transmit the stabilizing force from the substratum to the creeping soil. The lateral load on the dowels increases linearly up to a maximum value with displacement relative to the surrounding soil.

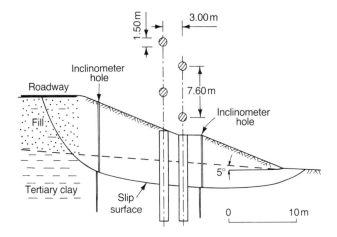

Figure 7.16 Dowelling in soil (after Gudehus and Schwarz, 1985)

7.3 SOIL ANCHORS

Soil anchors can be regarded as a form of tension pile comprising a rod
or cable grouted into a borehole and stressed after the grout has set. In
the case of soft ground excavations, they are used mainly as structural
'ties'. For example, they can be used to support diaphragm or sheet-pile
walls (Littlejohn, 1990). They can also be used to stabilize slopes, and to
strengthen and consolidate soil.

The capacity or pull-out resistance of a grouted anchor depends on the

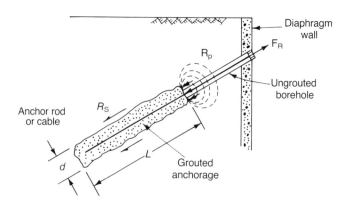

Figure 7.17 A stressed soil anchor. F_R = pull-out resistance, R_S = skin friction
resistance, R_P = end-bearing resistance, L = length of anchorage, d = diameter of
anchorage

amount of skin friction resistance and end-bearing resistance developed by the anchor as it is loaded (Figure 7.17). In the case of granular soils this resistance also depends on the depth, density and value of the angle of shearing resistance of the soil, as well as on the anchor dimensions. The size of the diameter of the anchorage depends on the type of soil and the method of construction. Cement grout is used for anchor construction and can limit the size of the diameter of the anchor in many soils. For instance, only in coarse sands and gravels are cement particles able to permeate the soil, radial penetration of grout forming a large anchor from a small-diameter borehole (Figure 7.18(a)). Resistance to withdrawal is governed principally by side shear but an end-bearing component may be included when calculating the pull-out capacity. Anchors in such ground are designed to resist safe working loads of 80 t (Littlejohn, 1970). The construction of the anchorage is carried out by drilling and driving a casing (102 mm nominal diameter) to the required depth, lowering a prepared steel cable or bar and then injecting grout under a nominal pressure (30–1000 kN/m^2) as the casing is withdrawn over the fixed anchor length.

In low-density soils, a larger anchor may be obtained by compaction of the soil. For example, gravel can be placed in the borehole and a small casing, fitted with a non-recoverable point is then driven into the gravel,

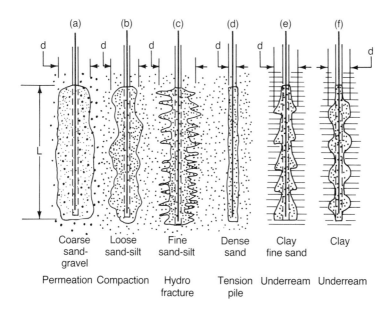

Figure 7.18 Various types of soil anchorages L = length of anchorage, d = diameter of anchorage

forcing it into the surrounding soil. The process enhances adhesion and produces a larger, uneven anchor, even if penetration is limited (Figure 7.18(b)). In addition, the grout also increases the density and angle of shearing resistance of the soil as well as forming an anchorage. Safe working loads of up to 30 t have been mobilized with this type of anchor (Littlejohn, 1970).

In stronger soils the ground may be subjected to hydrofracture. A packer is used in order to control hydrofracture in a lateral direction and so increase the diameter of the anchorage by radial fissuring (Figure 7.18(c)). Where stage grouting along a fixed anchor or re-grouting are envisaged, a *tube-à-manchette* can be employed. High-pressure grouted anchors are used mainly in cohesionless soils, although some success also has been obtained in stiff cohesive deposits.

When permeation, compaction or hydrofracture are not acceptable, tension piles (Figure 7.18(d)) or underreaming (Figure 7.18(e) and (f)) can be used. With tension piles, a large-diameter borehole or an extremely long anchor may be required to obtain a realistic anchor resistance, but in dense, deep sands, sufficient resistance may be developed by a 100 or 150 mm diameter anchor. Underreaming in sands and underreaming or 'belling' in clays spreads the anchor load and effectively increases the diameter of the anchorage. Indeed, in the case of clay anchors, underreaming is normally considered essential because of reduced adhesion at the clay–anchor interface due to softening of the clay in contact with drilling water or bleed water from grouting. Because clay is softened by water, the time taken for drilling, underreaming and grouting operations should be kept to a minimum. Wherever possible drilling for and grouting the fixed anchor should be done on the same day. In fact, when working in stiff-fissured clays a delay of 6 hours could be critical due to the deterioration of the borehole. An expanding underreamer is used to form a bell or a series of bells in the fixed anchor zone of the augered borehole, the number being influenced by the strength of the clay, after which the cable or rod is homed and grouted. Resistance to withdrawal is dependent primarily on side shear with an end-bearing component, although for single or widely spaced underreams the ground restraint may be mobilized chiefly by end bearing. This type of anchor mobilizes the full undrained shear strength of the clay and can resist safe working loads of 60 t. However, flushing water can penetrate fissures in clay during underreaming and initially may reduce the value of cohesion by half. Underreamed anchors must have their bells at least 5–6 m beyond any assumed slip plane.

Underreaming is ideally suited to clays with an undrained shear strength greater than 90 kN/m^2. Some difficulties, such as local collapse or breakdown of the neck portion between the underreams, are likely to occur when the undrained shear strength varies between 60 and 70 kN/m². At less than 50 kN/m² undrained shear strength, underreaming becomes virtually impossible. The use of high-pressure grouted anchors may be tried in such circumstances.

8

Geosynthetic materials

Geosynthetics are relatively thin, flexible polymeric materials. Over the past 25 years there has been a tremendous increase in their use due to the development of a large range of new materials which possess very different mechanical properties from those available previously (Giroud, 1986). When these materials are included in soil they improve its engineering performance and also lower the cost of construction.

8.1 TYPES OF GEOSYNTHETIC MATERIALS

Two main types of geosynthetics are used in geotechnical engineering: that is, woven and non-woven material. Usually they are composed of a single man-made material or a combination of two. The most widely used materials are polyamide (nylon), polypropylene, polyester (terylene) and polyethylene. Polyvinylidene chloride (PVC) is also used.

Nylon is a tough material, possessing good all-round properties (Table 8.1). It is available in an extensive range of fabric forms. However, it may undergo a slight loss in strength when soaked in water. Polyester generally offers a high resistance to breaking and tends to be the strongest of the more common polymers and the least extensible of the materials used in these fabrics. It also has good acid, abrasion and ultraviolet light resistance. Polyethylene combines high strength against breakage, good durability and resistance to breakdown by animals, fungus and chemical attack. It has a low density and it produces a strong, light fabric. Polypropylene is one of the weaker and more extensible synthetic fibres, and has a tendency to creep under constant loading. Polyvinylidene chloride has a high abrasion, good chemical resistance to chemicals and resists ultraviolet light.

Two main types of woven fabrics are available, namely, those manufactured from extruded filaments and those made from split film tapes.

Table 8.1 Mechanical properties, fabric form and resistance to attack of synthetic fibres used for geotextiles (after Cannon, 1976).

Property	Terylene	Nylon 66	Nylon 6	Nomex	Poly-ethylene	Poly-propylene	Teflon	Polyvinyl chloride	Viscous rayons	Jute/Wire
Fibre properties:										
Tenacity (grams/denier) app.	7·8	8	8	5·3	4·5	8	1·4	1·8	2·2	–
Extension at break (%) app.	9	15	17	22	25	18	15	25	18	–
Specific gravity	1·38	1·14	1·14	1·38	0·94	0·91	2·1	1·69	1·5	1·5
Melting point (°C)	260	250	215	370	120	165	327	–	190	–
Max. operating temp. (°C) app.	150	90	<65	200	55	90	280	–	<150	<65
Fabric form*:										
Woven	Yes	Yes	Yes	Yes	Yes	Yes	Yes	Yes	Yes	Yes
Knitted	Yes	Yes	Yes	Yes	Yes	–	Yes	Yes	Yes	Yes
Melded	Yes	Yes	Yes	–	Yes	Yes	–	Yes	No	No
Spun bonded	–	Yes	Yes	–	–	–	Yes	Yes	–	No
Needlefelt	Yes	Yes	Yes	Yes	Yes	Yes	Yes	Yes	Yes	Yes
Resistance to†:										
Fungus	1	3	3	3	4	3	3	3	3	1
Insects	2	2	2	2	4	2	3	3	2	1
Vermin	2	2	2	2	4	2	3	3	3	1
Mineral acids	3	2	2	2	4	4	4	3	1	1
Alkalis	2	3	3	3	4	4	4	3	4	1
Dry heat	3	2	2	4	2	2	4	2	2	2
Moist heat	2	3	3	3	2	2	4	2	1	2
Oxidizing agents	3	2	2	3	1	3	4	2	1	–
Abrasion	4	4	4	3	3	3	3	4	3	3
Ultraviolet light	4	3	3	3	1	3	4	4	2	1

* *Note:* Durability, long-term filtration properties, particle retention, water/liquid permeability, thickness and weight are all properties which depend upon the fabric construction process and have to be considered where appropriate.

† *Note:* 1, poor; 2, fair; 3, good; 4, excellent.

Extruded filaments have a circular cross-section and are stronger but more expensive to produce, weight for weight, than flat types. All woven fabrics are characterized by high strength and low breaking strains – that is, they possess low extensibility. This is not necessarily an advantage in soft ground where more extensible fabrics may prove more successful. Woven fabrics exhibit a slightly higher strength at 45° to the warp and weft directions than along them. In addition, they possess up to double the extensibility at 45°, than parallel to warp–weft directions. Unfortunately, if a woven fabric is torn, the tear can develop very rapidly at low applied stresses. However, woven fabrics made from multi-filament threads use several strands and so offer good levels of resistance to tearing.

Rankilor (1981) subdivided non-woven fabrics into two types, namely, thin and thick. The former are commonly referred to as two-dimensional fabrics, whereas the thicker fabrics are termed three-dimensional fabrics, or felts.

Two-dimensional fabrics have a uniform strength in all directions and possess a wide range of pore sizes. However, they generally are not as strong as three-dimensional fabrics and they are more extensible. On the other hand, their two-dimensional structure means they are not as susceptible to clogging by particles of soil.

Non-woven fabrics are produced in three ways: by needle punching, by melt bonding and by resin bonding. In the needle-punching process a layer of randomly oriented continuous fibres is laid on a felting machine and barbed needles are driven repeatedly through the fibres. Hence, the fibres are mechanically interlocked, being held together by physical entanglement (McGown, 1976). Fabric weights of 300–1000 g/m^2 usually are necessary to give acceptable levels of fabric strength for civil engineering purposes. Generally needle-punched non-woven fabrics offer a high resistance to tearing as the stress causes the filaments to move towards the direction of the applied force.

Melt-bonded fabrics consist of fibres which are melt bonded together at cross-over points. Higher strength is achieved with this type of manufacture at lower fabric weights (70–300 g/m^2) than with needle-punched fabrics. Two methods of bonding are used, homofil bonding and heterofil bonding. In the former (which produces spun-bonded fabric) the filaments are composed of a single polymer type but some of the filaments have different melting characteristics. In the latter, some of the filaments are composed of two types of polymer that have different melting points (heterofilaments). Spun-bonded fabrics exhibit a range of pore sizes in an individual fabric and generally have a lower extensibility than other melt fabrics. Non-woven materials produced by the melt-bonded process are generally isotropic and offer a high mechanical resistance. However, when the filaments have tightened, the resistance to tearing may be low. Also, the rigid bond between filaments can prevent the diffusion of localized mechanical deformations.

Usually in resin-bonded fabrics acrylic resin is sprayed onto or impregnated into a fibrous web. After curing, strong bonds are formed between filaments.

Geogrids and geonets are formed by either punching holes in a thick extruded sheet followed by unidirectional or bidirectional drawing to form an orthogonal grid, or by joining thick strands of extruded filaments at required cross-over points. Polypropylene and high-density polyethylene are used to manufacture geogrids, which means that they can resist attack by acids, alkalis and salts encountered in soils. They have a very long life when incorporated with soils and can maintain their engineering properties for decades even when exposed to sunlight. Furthermore, the molecular orientation of the polymer which occurs during its manufacture gives a high directional strength, as well as a high elastic modulus which ensures mobilization of high tensile strength at low strains and dramatically reduces creep. Geowebs are formed by weaving thick strands together.

Geomembranes are non-porous sheets that have very low permeability. The sheets may be formed of polyethylene, or non-woven polypropylene may be bonded to polyethylene.

Mats are robust three-dimensional geosynthetics with high tensile strength which are used to drape over soil slopes. The mat provides an erosion-resistant cover to the slope.

Geocomposites consist of an inner, more porous material sandwiched between two fabrics (Anon., 1989(b)). They are used in increasing quantities to replace conventional drainage systems.

8.2 PROPERTIES OF GEOSYNTHETIC MATERIALS

Synthetic fibres have many excellent qualities and can be individually designed to cover a wide range of properties. For example, they can possess high strength; flexibility; ability to elongate without rupture under high loadings and repetitive stresses; stability in chemicals, groundwaters, effluents and soils; resistance to fungal attack; and high abrasion resistance (Table 8.1).

Geosynthetics are composed of polymers, and the influence that the polymers have on behaviour depends very much on the structure of the geosynthetic produced by the manufacturing process. In the case of woven products in which the filaments, yarns or tapes of polymer are aligned in the direction of applied load, the behaviour of the geosynthetic relates closely to that of the basic polymer. By contrast, where the fibres or filaments of the polymer are randomly laid during manufacture and linked by bonding or needle punching, the influence of the polymer on the behaviour of the geosynthetic is very much less. In other words, the polymer properties have little influence on the properties of the fabric.

Enclosure of woven geosynthetics within soils has little effect on their load–strain behaviour, whereas with non-wovens, particularly needle-punched products, there is a considerable change in their behaviour. This is principally due to the soil particles locking the fibres into position, so reducing the amount of straightening and slippage at cross-overs and ends. This effect depends on the type of soil in contact with the geosynthetic, with fine sands and gravels giving the most pro-nounced effects.

The critical properties of geosynthetics depend upon the function they are meant to serve (McGown *et al.*, 1982). For example, pore size and shape are critical factors when a fabric is being used for filtration. When used as reinforcement the overall load extension and surface friction properties are critical. The structural arrangement of many geotextiles is liable to change when subjected to compressive stresses over their surfaces. Generally, the properties of woven fabrics are highly aniso-tropic as compared with those of non-woven fabrics.

8.2.1 Physical properties

The specific weight of a fabric is a fundamental property, being related to its strength, deformability and durability.

The rigidity of the structure of a fabric is as important as its complexity as far as filtration is concerned. For instance, a fabric that is open in the unloaded and loaded states gives a measure of the rigidity of its structure. A thin simple fabric structure is suitable for most cases of static filtration, whereas a thick complex fabric is required for plane drainage.

The pore size distribution of fabrics does not provide a direct measure of their particle-retaining capacity. Additional factors such as the particle size distribution of the soil, the flow rate and flow pattern of ground-water and particle size distribution of the downstream soil, also in-fluence the particle-retention capabilities of a fabric in a particular application. Nonetheless, the pore size distribution of an unloaded fabric offers a means of comparing the absolute pore size and size distribution of fabrics (Figure 8.1). The range of porosity of non-woven fabrics is much greater than that of woven fabrics. A non-woven fabric may possess a high porosity; for example, it may exceed 90% in the unloaded state or be 70–80% when loaded. The nature of the changes in the pore space of geotextiles vary with the level of confinement and the character of the material with which they are in contact. Changes in pore space alter the particle-retention capacity and permeability of the fabric.

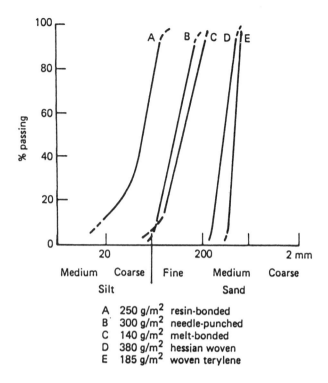

Figure 8.1 Typical fabric pore size distributions (after McGown, 1976)

The permeability of a fabric, depending on type of fabric used and structural arrangement, may vary by more than a factor of 10.

The hydraulic conductivities along and across fabric are significant characteristics in many applications. In order to gauge the influence of changes in fabric structure during loading, values of fabric permeability should be known for the unloaded and loaded conditions.

8.2.2 Mechanical properties

The stress–strain behaviour of geosynthetics exhibits an extremely wide range – for instance, some woven fabrics have strengths, moduli, break strains and creep properties which approach that of steel mesh (Figure 8.2). These woven fabrics, therefore, can be used instead of steel or aluminium strips in earth reinforcement. On the other hand, some non-wovens have relatively low strengths with breaking strains approaching 150% and high creep tendencies (Figure 8.3). The stress–strain characteristics of fabrics provide an indication of their relative toughness. In situations where large strains have to be accepted, extensible non-

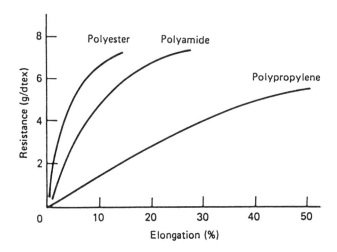

Figure 8.2   Stress–strain curves for various filaments (from Hoare, 1987)

woven fabrics prove suitable. Conversely, the use of a stiffer woven fabric is necessary when strains in the soil must be limited.

Generally, geosynthetics are more extensible than the soils in which they are placed in that they need a greater amount of tensile strain to cause rupture than soils require to achieve peak strength (McGown *et al.*, 1982). Hence, the tensile resistance that can be utilized in geosynthetics is very much less than their rupture strength. In fact, in most cases the strength that can be utilized is limited by the permissible strains in the soil, which rarely exceed 10% and in many applications are very much less. Consequently, as far as soil reinforcement applications are concerned, it is the load–strain–time behaviour of geosynthetics up to 10% strain that is important.

The peak tensile strength and strain measured under plane strain conditions offers an indication of the maximum load capacity of a fabric and the strain required to achieve this. Woven fabrics, on a weight for weight basis, afford much higher resistance to breaking than non-woven types. The latter generally are characterized by relatively low strength and high failure elongation. In general, the percentage loss in strength is much less for non-woven than for woven fabrics. However, the higher initial strength of woven fabrics must be borne in mind. The structural arrangement of a fabric may give rise to various degrees of anisotropy in terms of strength and extensibility. Changes in the internal structural mobility, interfibre friction and the surface texture of geotextiles are likely to occur according to the level and nature of confinement. In this way the load-extension and surface friction characteristics of a fabric are altered. In fact, Ingold (1991) referred to a transition point at which the

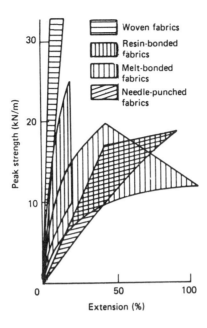

Figure 8.3   Typical plane strain data for different types of geotextiles, normalized to 300 g/m³ fabric weight (after McGown, 1976)

mechanism controlling tensile failure underwent a change, the time at which this occurred being affected by polymer type, environmental conditions and any damage caused by construction operations. Because the tensile strengths of different geosynthetic products differ and are affected to differing degrees in different environments, the new British code of practice for reinforced soils (Anon., 1991) has adopted the concept of partial factors of safety, using a partial load factor and a partial material factor to obtain a margin of safety against failure.

In certain situations, especially in extensible strain reinforcement situations where the inclusions redistribute the strains within the soil mass, loads are applied unevenly within the fabric. The 'grab' tensile test can be used to assess localized loading and takes account of any contribution made by the fabric that is not in the direct stress path between loaded points. This contribution can be large for some materials, particularly non-wovens with highly random arrangements of fibres. Typical values for fabrics are given in Table 8.2.

The bursting load of fabric and extension to burst are important properties. Resistance to small deformations that are likely to occur on site can be measured with a Mullen burst test. Values of bursting load obtained by such tests are given in Table 8.3. Tear propagation, rather

Table 8.2 Typical grab tensile characteristics of some fabric types used in civil engineering (normalised to 200g/m²) (after McKeand and Sissons, 1978)

	Breaking load (N)	Extension to break (%)	Load at 5% extn. (N)	Rupture energy (Nm)
Melt-bonded: Melded	900–1400	70–125	110–240	50–90
Other	1000–1200	50–60	250–300	40–50
Needle-punched	750–950	65–150	10–20	20–65
Resin-bonded	650–800	50–65	100–250	20–25
Woven tape	800–1000	10–25	300–350	10–25

than tear initiation, is another important property. In some applications such as river bank and coastal protection, this may be one of the major requirements of the fabrics used. Also they should not be affected significantly by punching loads.

When loads have to be sustained continuously – as, for example, in earth reinforcement – the creep-relaxation properties of the fabrics are of major consequence. Creep is a function not only of the type of material, the production method and form of reinforcement, but also of the level of stress imposed, the length of time involved, environmental attack and the operational temperature. Creep can result in rupture at lower than the recognized breaking stress and in order to apply satisfactory factors of safety, the load placed on fabric should be limited to less than 50% of the measured breaking load. Generally, the load at which tensile rupture occurs, at a given temperature, decreases with time. Creep coefficients for polyester are lower for lower levels of loading, though this is not the case for nylon yarn. Creep in polypropylene fibre is generally of a higher order. When a fabric is subjected to tension over a

Table 8.3 Typical values of Mullen burst for some fabric types used in civil engineering (normalised to 200g/m²) (after McKeand and Sissons, 1978)

	Bursting load (N/cm²)	Height of dome at burst (cm)
Melt-bonded: Melded	100–200	1.4–2.1
Other	150–200	1.4–1.6
Needle-punched	100–210	1.1–1.9
Resin-bonded	150–270	0.9–1.1
Woven tape	350–380	1.0–1.2

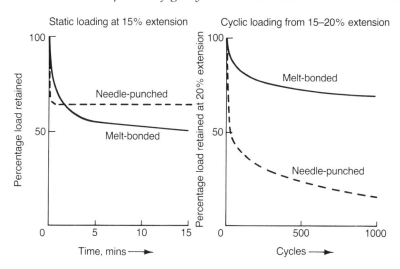

Figure 8.4 Stress decay on static and cyclic loading of non-woven fabrics (after McKeand and Sissons, 1978)

period of time, stress decay measurements provide an indication of creep behaviour. Such measurements show that needle-punched non-woven fabrics exhibit better stress decay characteristics than melt-bonded materials under static loading. The situation is dramatically reversed under cyclic loading (Figure 8.4).

The transfer of load from soil to fabric and between overlapped layers of fabric is dependent on friction. Hence, the coefficients of soil–fabric friction and fabric–fabric friction are important properties. As the fabric is subject to frictional contacts, the abrasion resistance of the fabric is also important.

8.2.3 Environmental properties

Corrosion of geosynthetics can take place due to the presence of certain chemicals in substantial amounts in aggressive acid or alkaline soils. For example, the presence of sulphuric acid, together with the activity of sulphur bacteria, can bring about deterioration of nylon-based products. Polyamides are attacked by strong acids and polyesters are susceptible to strong alkalis, but both have excellent resistance over the pH range (3–10) likely to be encountered in most soils. Oxidation can cause degradation of nylon, polyesters and polyolefines. Water is absorbed by and diffused through most polymers at ambient temperatures and humidities. Indeed, in composite materials the possibility of preferential diffusion of water exists, diffusion occurring due to capillary action along, for example, interlaminar planes.

Some micro-organisms present in soils are detrimental to synthetic products. Microbial activity is most significant in humid tropical areas.

It appears that the tensile strength of some types of nylon deteriorates with time after incorporation in soil. In some cases the material has been found to become brittle, rupture then occurring without any appreciable elongation. This may be attributable to hydrolysis. The strength of polyester, polythene and polypropylene undergo negligible or insignificant changes after burial.

During construction, exposure of geotextiles to some direct sunlight and other weather effects is unavoidable, the effects of exposure being especially important in hot climates. Hence, a factor that must be considered when geotextiles are used in construction is their degradation when exposed to weather and ultraviolet light. The effects of weathering on geotextiles depends on their geographical location. For example, Brand and Pang (1991) quoted tests carried out in Florida, North Carolina and Arizona on the one hand, and Sweden and England on the other. In the North American tests some geotextiles subjected to exposure had a half-life (time to 50% strength loss) of less than 8 weeks, while the half-life of others exceeded 32 weeks. The woven polypropylene geotextile exposed in England and Sweden underwent strength losses of 14–22% in a year compared with losses of 2–5% per annum for buried samples. The prediction of the performance of a geotextile upon exposure should be based on expected sun-hours; the energy of sunlight falling on the geotextile; temperature, humidity and atmosphere pollution; and the temperature of the geotextile. Moreover, different polymers have different sensitivities (e.g. polyester is sensitive to wavelengths in the region of 325 nm, polyethylene to 300 nm and polypropylene to 370 nm). The weakening of synthetic polymers by ultraviolent light can be partially compensated for by addition of an UV stabilizing agent (Cooke and Rebenfeld, 1988).

A series of natural exposure tests carried out in Hong Kong by Brand and Pang (1991) showed that all the geotextiles became brittle with a corresponding loss in tensile strength. The loss of strength in the first month was less than 16%, with some geotextiles performing better than others. The long-term performance of the geotextiles varied considerably, with the half-life ranging from 3 to 9 months. Some geotextiles virtually lost their strength in 6 months. The rate of strength loss generally increased with radiation intensity. Brand and Pang found that polypropylene geotextiles without UV stabilizers offered the lowest resistance to weathering, while needle-punched polyester geotextiles without UV stabilizers performed in a similar way to the UV stabilized polypropylene needle-punched geotextiles. Needle-punched geotextiles

($\geq$ 1.45 mm thick) gave better long-term performance than the thinner and more open-structured woven geotextiles, irrespective of polymer type. In the case of polypropylene geotextiles, needle-punched fabrics had better resistance to degradation than the thinner heat-bonded geotextiles. Accordingly, Brand and Pang recommended that rolls of geotextiles should remain covered until they are used and that they then should be buried as soon as possible.

8.2.4 Properties characteristic of geogrids

When placed in a soil the ribs on a geogrid which run transverse to the direction of primary loading provide a series of anchors (Jewell *et al.*, 1984(a)). Consequently, stress is transferred to the grid not just by surface friction, as with strip reinforcement, but also by interaction. This provides a very effective means of stress transfer which mobilizes the maximum benefit from the grid reinforcement and minimizes anchorage lengths. The open structure of a geogrid interlocks and interacts with the soil to provide a high resistance to sliding.

In many reinforcement applications, high modulus geogrids are better suited than low moduli geosynthetics since they mobilize reinforcement properties at elongations that are more compatible with working strains in soils. Typical compacted soils have maximum working strains around 1–4% and at elongations in this range only very high modulus geogrids can mobilize significant tensile forces.

8.3 FUNCTIONS OF GEOSYNTHETICS

Geosynthetics perform a number of basic functions, namely, separation (segregation of two layers of solid particles); cushioning (ability to absorb impact or abrasion from impinging materials); filtration (passage of liquid/gas across an interface with retention of solids); reinforcement (redistribution of the stress–strain pattern in a load-carrying system); and, in the case of thicker fabrics, drainage in the plane of the fabric (McGown, 1976; Fluet, 1988). In some applications one function may be dominant. However, many applications of a fabric involve a combination of more than one of the basic functions. Geomembranes serve an isolation or protection function.

An example of the separation function is the separation of the aggregate of a roadbed from the subgrade below. The geotextile prevents the loss of aggregate to the subgrade while, at the same time, preventing the subgrade material from being intruded into the aggregate. Thick non-woven fabrics can perform a cushion function, protecting one material from damage by another. An example is provided by the use of non-woven geotextiles in liners to protect geomembranes from overlying gravel. In the filtration function a geotextile retains soil while allowing the passage of water. This occurs when filters are used for erosion control as well as drainage. The purpose of a geosynthetic material when used for reinforcement is to add tensile strength to the soil and thus construct an earth structure with adequate compressive strength (derived from the soil) and tensile strength. Geogrids provide higher low strain elastic moduli than other commercially available geosynthetics, while specialized woven geotextiles provide the highest tensile strengths. Lateral drainage or transmission of fluid through the plane of the geosynthetic requires a material of adequate thickness and permeability. In addition, the material must possess sufficient dimensional stability to retain its thickness under pressure. Geomembranes, because of their impermeability, can be used to isolate and retain fluids. For example, geomembranes may be used to provide isolation (i.e. waterproofing) in tunnel construction or to protect foundations in aggressive soils.

The maximum pore size of a fabric represents the largest particle size that can pass through a fabric without tearing it. However, in most situations this particle size is retained. Thus the particle-retention capacity and pore size distribution of fabrics are not the same.

The particle-retention capacity of a fabric is not a constant. It depends on the characteristics of the materials in contact with it as well as the hydraulic conditions of the soil in which it is incorporated. Where non-reversing flow occurs, the particle size distribution of the downstream soil can reduce the effective size of the pore opening. In addition, if the upstream soil is well graded, bridging of particles may occur across the pore openings which appreciably reduces the likelihood of many particle sizes moving through the fabric. The higher the tortuosity of the pores within the fabric, the more effective is the retention capacity. In reversing flow conditions the bridging effect may not develop and particles may be moved in and out of the fabric. This tends to increase the size of particle migration over that in non-reversing flow conditions. As the pore size distribution of a fabric may be altered significantly when loaded, it is necessary when considering the relative usefulness of fabrics performing as separators, filters, drains or reinforcements to test the fabrics in the soil (McGown *et al.*, 1978). As far as separation is concerned, most fabrics perform satisfactorily under non-reversing and low hydrodynamic reversing flow conditions. Where the dynamic effects are significant the range of applicability of fabrics can be appreciably increased by using them in conjunction with a layer of filter sand.

Filtration and drainage-in-the-plane functions occur over a range of flow conditions, including steady state and dissipating head, non-reversing flow and variable head-reversing flow. Hoare (1978) noted that thin fabrics are adequate for non-reversing flow conditions while thick compressible fabrics can be used for hydrodynamic reversing flow conditions. Alternatively, a thin fabric together with some granular material can be used to replace one layer of a multi-layer filter.

Small particles of soil immediately in contact with a fabric are the first to move through it, creating voids that may be occupied or bridged by other soil particles. Migration and re-orientation of soil particles continues and eventually creates an upstream filter within the soil (see Figure 4.1). The development of this upstream filter depends on the size and number of particles that are involved in migration and the hydraulic conductivity of the screen. The effectiveness of the screen, in turn, is governed by the structure of the fabric, the properties of the upstream soil, the drainage media grading curve and the fluid flow conditions (McGown *et al.*, 1978).

Woven fabrics and most melt resin-bonded non-woven fabrics do not possess sufficient hydraulic conductivity in the plane of the fabric to act as drains. However, the thicker needle-punched non-wovens, as a consequence of their bulk, offer some flow capacity. A drain is formed by enclosing granular filter material in fabric (see Chapter 4). Fabric-wrapped drains provide greater consistency in filter design, installation and performance than conventional drains. Purpose-made prefabricated geotextiles which contain a core of channels enclosed by fabric on either side are now used for drainage purposes. The use of purpose-made drainage fabrics eliminates the use of granular filler material, an important consideration when suitable granular material is not readily available.

The purpose of placing tension-resistant members in soil structures is to increase the resistance of these structures to shear failure. The tensile strength of a fabric affords tensile strength to an earth mass (Hoare, 1978), the tensile strength of the fabric being used to resist forces parallel to the plane of the fabric as in earth-reinforced walls. In addition, the shearing resistance of the soil is increased. The basic operational mechanism is strain controlled (Juran *et al.*, 1990(a)). In fact, tension can only develop in reinforcing members when incremental tensile strains are developed in the soil and in the planes of the reinforcing members. Therefore, no matter what the level of stress, if tensile strains cannot occur within the soil and in the planes of the reinforcements, then tension will not be developed in the reinforcements. The inclusion of fabric within an earth mass redistributes stress in each membrane. Not only are stress patterns altered when sheet fabrics are used for reinforcement, but strain directions also are changed substantially. When strips of webbing are used in a reinforced soil system, they more or less act as passive tie-backs. They are attached to the facing units.

As mentioned above, most non-woven fabrics have elastic moduli much less than that of the soil, and can strain far beyond the peak strain of the soil. Consequently, they have a very low load take-off capacity in a fabric-reinforced system. Nevertheless, the presence of such fabrics in soil enhances its load-carrying capacity. Where strains in a system must be restricted, then geosynthetics with the highest values of strength and deformation moduli, and the lowest break strains, should be used. Conversely, in systems where strain can be increased significantly, the inclusion of fabrics can lead to more ductile, less brittle system be- haviour. This, in turn, can give rise to greater load-bearing capacity. A good example of application of the former is in reinforced earth retaining walls and of the latter is in granular soil embankments over highly compressible soils.

8.4 USES OF GEOSYNTHETIC MATERIALS

The improvement in the performance of a pavement attributable to the inclusion of fabric was one of the first ways in which geotextiles were employed in civil engineering and remains the most important use (Anon., 1990(b)). The improvement comes mainly from the separation and reinforcing functions, and can be assessed in terms of either an improved system performance (e.g. reduction in deformation or in- crease in traffic passes before failure) or reduced aggregate thickness requirements (where reductions of the order 25–50% are feasible for low-strength subgrade conditions with suitable geosynthetics).

The most frequent role of fabric in road construction is as a separator between the sub-base and subgrade. This prevents the subgrade mat- erial from intruding into the sub-base due to repeated traffic loading and so increases the bearing capacity of the system. The savings in sub-base materials, which would otherwise be lost due to mixing with the subgrade, can sometimes cover the cost of the fabric. The range of gradings or materials that can be used as sub-bases with fabrics is normally greater than when fabrics are not used. Nevertheless, the sub- base materials preferably should be angular, compactible and suffi- ciently well graded to provide a good riding surface.

The position of a fabric in a system is critical and the use of two discreet layers can afford much greater improvement than just twice that of a single layer. McGown (1976) developed a series of fabric- included designs which suggest modes of fabric inclusion that give increasing degrees of improvement, depending on subgrade and sub- base types.

If a fabric is to increase the bearing capacity of a subsoil or pavement significantly, then large deformations of the soil–fabric system generally must be accepted as the fabric has no bending stiffness, is relatively extensible, is usually laid horizontally and is normally restrained from extending laterally. Thus, considerable vertical movement is required to provide the necessary stretching to induce the tension that affords vertical load-carrying capacity to the fabric. Fabrics, therefore, are likely to be of most use when included within low-density sands and very soft clays. Although such large deformations may be acceptable for access and haul roads they are not acceptable for most permanent pavements. In this case the geosynthetic at the sub-base/subgrade interface should not be subjected to mechanical stress or abrasion. When fabric is used in temporary or permanent road construction it helps redistribute the load above any local soft spots that occur in a subgrade. In other words, the fabric deforms locally and progressively redistributes load to the surrounding areas thereby limiting local deflections. As a result, the extent of the severity of local pavement failure and differential settlement is reduced.

When fabric is used for unbound temporary access or haul roads, it is placed on the subgrade and then overlain with aggregate (Figure 8.5). Ground penetration by aggregate is minimal and the extensibility of the fabric allows it to conform to the surface over which it is laid. The fabric also prevents sub-base deterioration by reducing pumping, whereby fines become suspended in water and are transported into the sub-base by the action of wheel loading. The incorporation of fabric at the sub-base/subgrade interface constrains the sub-base and therefore reduces the development of ruts in the road. The subgrade material undergoes a long-term loss of moisture in its upper layers, which further increases its strength.

When constructing roads, railways or runways over variable sub-grades, geogrids can be used within a granular capping layer. To achieve maximum reinforcing effect, the vertical spacing between grid layers should not be greater than 500 mm. Multiple layer systems are possible since, by interlocking with sub-base particles, the grid does not create a weak shear plane.

The use of geogrid reinforcement in road construction helps restrain lateral expansive movements at the base of the aggregate under wheel loading. This gives rise to an improved load redistribution in the sub-base which, in turn, means a reduced pressure on the subgrade. In addition, the cyclic strains in the pavement are reduced. The stiff load-bearing platform created by the interlocking of granular fill with geogrids is utilized effectively in the construction of roads over weak soil. Reduction in the required aggregate thickness of 30–50% may be achieved. Geogrids have been used to construct access roads across peat bogs, the geogrid enabling the roads to be 'floated' over the surface.

In arid regions geomembrane can be used as capillary breaks to stop the upward movement of salts where they would destroy the road surface. Geomembranes also can be used to prevent the formation of ice

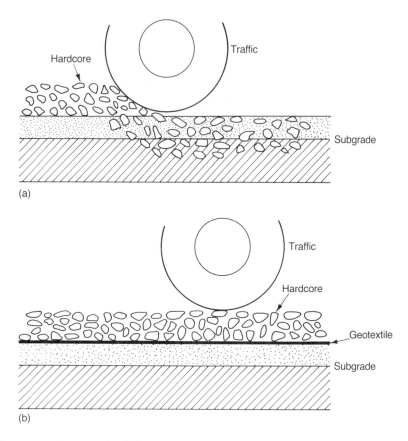

Figure 8.5 (a) Loss of sub-base material due to a reduction in subgrade strength; and (b) use of geotextile to solve the problem

lenses in permafrost regions. The membrane must be located below the frost line and above the water table.

Where there is a likelihood of uplift pressure disturbing a road constructed below the piezometric level, it is important to install a horizontal drainage blanket. This intercepts the rising water and conveys it laterally to drains at the side of the road. A geocomposite can be used for effective horizontal drainage.

Problems may arise when sub-bases are used that are sensitive to moisture changes – that is, they swell, shrink or degrade. In such instances it is best to envelop the sub-base, or excavate, replace and compact the upper layers of sub-base in an envelope of impermeable membrane. A membrane of limited permeability is acceptable provided that any changes in moisture content in the encapsulated soil does not result in a critical loss of strength during periods of heavy rainfall or flooding, such as experienced in semi-arid regions. Soil that is on the dry side of optimum moisture content is favoured for a membrane-encapsulated soil layer (MESL), since higher strengths can be obtained, especially in soil with a high percentage of fines. The MESL brings about a better distribution of stress in a road pavement than if a membrane alone was used, and has been used for rapidly constructed road and airfield pavements (Lawson and Ingles, 1982).

Geosynthetic material with high tensile strength contributes to the load-carrying capacity of soil that is poor in tension, and is used as reinforcement in the construction of earth-reinforced walls and embankments (Ingold, 1984(b); Leshchinsky and Boedeker, 1989). The reinforcement develops tensile forces which increase the shearing resistance of the soil and improve stability. The effectiveness of the reinforcement is principally a function of its strength, stiffness, location within the soil mass and its ability to bond with soil. The force in the reinforcement may be resolved perpendicular and parallel to the inclination of the failure surface. The component of the force perpendicular to the failure surface acts with the angle of friction of the soil to increase the frictional resistance. The component parallel to the failure surface resists the shear forces along the surface.

A wide variety of fill materials and facings can be used with geogrids in the construction of reinforced soil walls. Horizontally laid layers of geogrid provide the structural stability while the facing contains the fill and provides a suitable aesthetic appearance. Reinforced soil structures have a high tolerance of differential settlement thereby avoiding the requirements for expensive foundations. Wall facings can consist of concrete panels, timber, brick, stone or gabions. Alternatively, soft facings can be formed by wrapping the grid layers up the face around successive lifts. Such faces can be detailed to provide vegetation.

In reinforced earth structures, there are three methods of construction which accommodate settlements that occur within the soil mass: namely, the concertina method, the telescopic method and the sliding method. In the concertina method the face is formed by wrapping the reinforcement (geogrid) around the compacted soil layer so that successive layers of fill are completely enclosed (Figure 8.6). The exposed geosynthetic material is covered with asphalt emulsion, shotcrete or hydroseeded soil for long-term protection against ultraviolet light and weathering. Deformation of the face accommodates the settlement. In the telescopic method a short section of geogrid is cast into the facing panels and joined to the reinforcing grid by a dowel bar. Construction settlements within the soil mass in the telescopic method are accommodated by the facing panels closing up. A gap is left between the panels when they are attached to the reinforcing elements. Settlement is accommodated in the sliding method by allowing some freedom of the

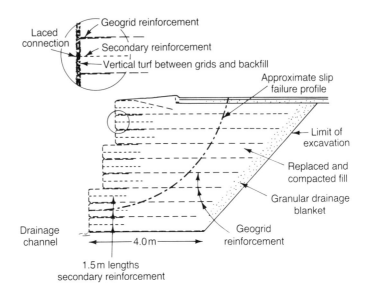

Laced connection

Geogrid reinforcement

Secondary reinforcement

Vertical turf between grids and backfill

Approximate slip failure profile

Limit of excavation

Replaced and compacted fill

Granular drainage blanket

Drainage channel

Geogrid reinforcement

4.0 m

1.5 m lengths secondary reinforcement

Figure 8.6 The concertina method of construction of retaining walls involves forming the face by returning the reinforcement over the compacted layer

reinforcement in the soil to slide down the facing while remaining connected to it.

Rising land prices, the scarcity of good quality fill and the need to widen existing highways provide incentives to steepen slopes and to utilize marginal fills. In such cases the reinforcement is laid directly on the surface of a layer of compacted fill and covered with the succeeding layer of fill. Geogrids may be used to steepen embankment slopes up to 90°. A face support is required for slope angles between 45 and 90°. The geogrids can be wrapped around successive lifts of fill, removing the need for rigid facings. Generally geogrid-reinforced slopes shallower than 45° do not require the wrap-around detail to provide face stability. In such instances a mat can be laid over the face, prior to placement of topsoil, to resist erosion and accelerate and promote the growth of vegetation (Figure 8.7).

Geogrids or geomats can be used in the construction of embankments over poor ground without the need to excavate the ground and substitute granular fill. They can allow acceleration of fill placement often in conjunction with vertical band drains. Layers of geogrid or geowebs can be used at the base of the embankment to intersect potential deep failure surfaces. Rowe *et al.* (1984) described the use of strong geotextiles to help construct an embankment over peat deposits. Geogrids can also be used to encapsulate a drainage layer of granular

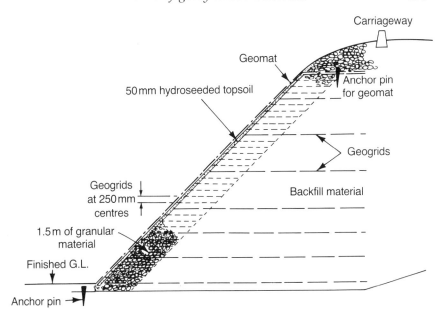

Figure 8.7   Reinforced soil embankment

material at the base of the embankment. Both of these methods help reduce and regulate differential settlement. A geocell mattress can also be constructed at the base of an embankment that is to be constructed on soft soil. The cells are generally about 1 m high and are filled with granular material (Figure 8.8). This also acts as a drainage layer. The mattress intersects potential failure planes and its rigidity forces them deeper into firmer soil. The rough interface at the base of the mattress

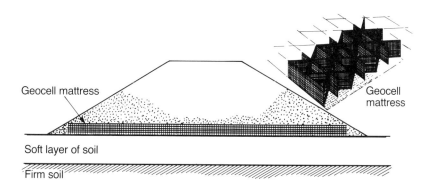

Figure 8.8   Geocell mattress beneath embankment

ensures mobilization of the maximum shear capacity of the foundation soil and significantly increases stability. Differential settlement and lateral spread are minimized.

Traditionally, slips have been reinstated by excavation and disposal of the failed soil and substitution with a free-draining granular fill. Re-use of the slipped soil, reinforced with geogrids to intersect potential failure surfaces, has enabled repairs to be carried out at a quarter of the traditional cost. Geogrids are usually installed by excavating the failed slope in sections. Excavation is continued beyond the slip surface to provide sufficient anchorage length for the layers of geogrid reinforcement. The removed material is then used to reinstate the slope. Horizontal layers of reinforcement are incorporated at appropriate vertical spacings to increase the factor of safety and prevent further slip failure. The procedure is repeated until the whole failed slope has been repaired in a series of sections. Granular drainage layers may be incorporated within the repair to further enhance stability. The addition of lime aids handling of particularly wet clay without affecting the performance of the geogrid.

Geosynthetics can be used when reclaiming poor-quality low-lying land. The geosynthetic is rolled directly over the ground surface if a permeable fill, such as sand, is to be laid above. However, if the only material available is impermeable, then a drainage blanket should be positioned beneath the fabric to convey away the groundwater. Two or more layers of fabric, each pair separated by a thin layer of fill, can be used on very soft ground that has a high moisture content. Improvements of soft ground with geogrids and geomeshes have been described by Kamauchi and Kitamori (1985) and Watari *et al.* (1986).

Geotextiles also have been used to reclaim mine or quarry tailings ponds. The whole surface is covered with a strong permeable membrane which is then overlain by a thin layer of compacted granular fill. Broms and Shirlaw (1987) described the use of geotextile to increase the bearing capacity of very soft tailings in waste ponds so that a high fill, required for preloading, could be constructed. The fabric was first placed over the area to be stabilized, after which berms were placed around the perimeter of the area to hold the fabric in place. Narrow berms were then constructed across the fabric and then widened until the whole area was covered.

The use of geosynthetics as a filter media is next in importance to their use in pavement construction. The performance of a filter drain is governed by the properties of the soil in which it is incorporated, the hydraulic flow conditions and by the properties of the fabric (Ingold, 1984(a)).

A fabric filter brings about a 'self-induced' filter action within the soil immediately around it (Figure 4.1). In other words, some fines in the soil

Table 8.4 Criteria for preventing soil piping in one-way flow situations (after Giroud, 1982)

Relative density (D_R)	$1 < C_U < 3$	$C_U > 3$
Loose ($D_R < 35\%$)	$O_{95} < (C_U)(D_{50})^{\#}$	$O_{95} < (9/C_U)D_{50}$
Intermediate ($35\% < D_R < 65\%$)	$O_{95} < 1.5(C_U)(D_{50})$	$O_{95} < (13.5/C_U)D_{50}$
Dense ($D_R > 65\%$)	$O_{95} < 2(C_U)(D_{50})$	$O_{95} < (18/C_U)D_{50}$

$$D_R = \frac{e_{max} - e}{e_{max} - e_{min}}$$

where e is the naturally occurring void ratio, e_{max} is the maximum void ratio and e_{min} is the minimum void ratio.

$^{\#}O_{95}$ = apparent opening size of geosynthetic; unfortunately this is not easy to determine
D_{50} = particle size corresponding to 50% finer than maximum size
C_U = D_{60}/D_{10} = coefficient of uniformity

next to the fabric can pass through it and so be transported away while most remain in place. This process of filter development continues until a new equilibrium condition is established. From then on only clean water passes through the filter. The filter criteria of geotextiles can be related to the particle size characteristics of the soil concerned. For example, Giroud (1982) suggested a set of criteria to prevent piping (i.e. soil passing through the geotextile), which are given in Table 8.4.

Fabrics have been used to form various types of drains such as linear drains where fabric encapsulates granular material in a trench, drainage blankets beneath fills, or fin drains (see Chapter 4). In the latter case the geocomposite consists of a sandwich of thin, high-porosity filter fabric enclosing a core able to collect water. This runs to a perforated unplasticized PVC pipe around which it is wrapped (see Figure 4.3). The core is formed from an extruded polyethylene mesh sheet. Geocomposite is used to drain both sides of French drains, and in cut-off drainage where the core is sheathed on one side by filter fabric and impermeable geomembrane on the other.

Geomembranes are incorporated into clay blankets to form composite liners for waste disposal sites. Although resistant to many chemicals, some geomembranes are susceptible to degradation by organic solvents. Hence, the United States Environmental Protection Agency recommends the use of double liner systems (Figure 8.9).

Geosynthetics are used for erosion control, for example, to retain soil undergoing erosion, or in the construction of river and marine defence works. Geomats, as noted, are three-dimensional materials with high tensile strength and excellent drape qualities. They are used to protect slopes, the mat stabilizing the surface layer of soil, and when filled with soil help establish a vegetative cover on slopes. Mats are draped over the slope requiring protection and should maintain intimate contact with the soil beneath. Any existing vegetation should be cropped to ground

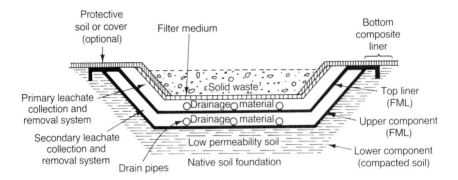

Figure 8.9 Double liner system containing two flexible membrane liners (FML) for disposal of hazardous waste

level or removed, and bumps and hollows levelled out. The mats are pegged to the soil, the pegs being inserted at the edges of the mats. The pegs vary in length between 100 and 300 mm. On steeper slopes the mats should be buried in shallow V-shaped trenches both at the toe and just beyond the crest of the slope. Any seeding or application of fertilizer to the soil obviously should be carried out prior to the mats being laid.

Biodegradable geotextiles – that is, prefabricated natural materials – are used to accelerate the establishment of vegetation on newly formed slopes, thereby increasing their erosion resistance and stability against shallow-seated slope failure (Barker, 1986). These geotextiles take the form of open mesh jute (or hessian) netting, and three-dimensional quilted mats consisting of coir (coconut fibre) and straw mixtures, or poplar or pine shavings. Biodegradable mats are essentially reinforced mulches, shielding soil from erosion by rain and insulating it from extremes of temperature and moisture. They can retain more than their own weight of water, thereby providing resistance to dry spells. In addition, biodegradable geotextiles can contain seeded mixes (e.g. grass or tree seeds mixed with cotton waste). Such materials protect the slope, retain soil moisture, enhance the humus content of the soil when they degrade and offer reinforcement to the turf that forms.

Geosynthetics used in river defence works usually require high strength. Consideration has to be given to the velocity of river water during placing, as well as to the type of soil being protected. Above water the fabric is laid over the soil; however, fabric placed below water usually needs to be weighed down. The rip-rap overlying the fabric should be placed at the toe first and then built up the slope. Webbing can be used directly for river defences without stone cover. Flood bunds can be reinforced with geogrids.

A fabric may be placed behind a heavy sea defence structure to act as a backing separator. In such a case it is necessary to dissipate uplift pressures and relieve groundwater flow without clogging occurring. A fabric beneath heavy armour also acts as a filter and may have to dissipate energy as well as withstand direct wave pressure and wave-induced movement of stones. Geogrids can also be used to form gabions and mattresses for cliff protection, sea walls or permeable groynes.

9

Grouts and grouting

9.1 INTRODUCTION

Grouting usually refers to the injection of suspensions, solutions and emulsions into pores in soils to improve their geotechnical characteristics. Hence permeation grouting is the commonest type of grouting used in construction. However, grout may be used to displace soils as in compaction and claquage (hydrofracture) grouting or to replace soils as in jet grouting. Grouting is widely used to reduce the permeability and/ or to increase the strength of soil. If grouting is to be effective, it must reliably penetrate the ground to the required distance and arrive in a satisfactory state.

Grouts can be grouped into two basic categories: suspension or particulate grouts (Bingham fluids); and solution or non-particulate grouts (Newtonian fluids). Particulate grouts consist of cement–water, clay–water or cement–clay–water mixes. The most common classes of chemical grouts are silicates, lignins, resins, acrylamides and urethanes (Karol, 1968; Chi and Yang, 1985). The silicate grouts are the most widely used chemical grouts (over 90% of present use).

Grout used to reduce the permeability of the ground must be able to develop sufficient strength to withstand the hydraulic gradient imposed. Normally a cement or clay–cement grout is used in coarser soils and clay–chemical or chemical grouts are used in finer grained soils (Figure 9.1). Microfine cement has been used to grout fine sands (Arenzana *et al.*, 1989). In the case of chemical grouts, especially on large jobs, cheaper high-viscosity grouts may be used to fill the larger voids and more costly low-viscosity grouts to fill voids too small to be penetrated by the initial grouting. While it is not possible to state with any certainty what will be the result of any given grouting programme, it is fairly certain that a given degree of watertightness can be achieved if sufficient grouting is performed. In general terms a cut-off with a

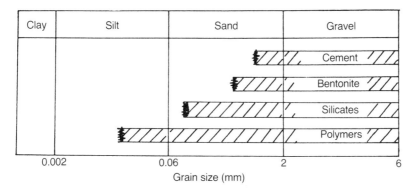

Figure 9.1 Soil size limitations on grout permeation (after Mitchell, 1970, with permission of ASCE)

permeability of 10^{-6} m/s in alluvium is attainable, although in individual cases an improvement of one or more orders of magnitude above this figure has been claimed.

Grouting can be used to strengthen the ground. For example, coarse-grained granular material can be consolidated to strengths similar to concrete when cement grout is injected. Indeed, when grouting to enhance strength, a cement-based grout is preferred if the particle size distribution of the soil permits, because this gives a higher strength compared with other grouts. If this is not the case, chemicals must be used since clay-based grouts add little strength to the soil. Resin- or polymer-based grouts are usually stronger than silicate grouts.

A site investigation should be undertaken prior to planning a grouting programme. The investigation should provide details of the different soil types that require grouting, together with their location and thickness. When the object of grouting is to reduce permeability, porosity and permeability data should be obtained, as should those relating to hydraulic gradients and the chemistry of the groundwater. The latter may influence the type of grout chosen and the extent of treatment, while porosity affects the amount of grout used. The co-efficient of permeability may set the lower limits for permeation grouting. Where ground strengthening is necessary, additional tests are required such as penetrometer or pressuremeter tests or laboratory tests on undisturbed samples to assess their strength.

The determination of the most suitable grout to be injected under given conditions requires data relating, firstly, to the size of the voids requiring grouting and, secondly, to the permeability of the ground. The arrangement of the soil strata to be grouted and the thickness of the overburden should also be known since these influence the maximum grout pressure that can be used. As far as the grout itself is concerned, it must be easy to mix and inject – that is, capable of flowing readily through the void space under reasonable pressure. It must form a substance with adequate strength when set so that it is not displaced, and it must set with a minimum amount of shrinkage and remain stable. The placement of grout should not contaminate the groundwater.

9.2 PROPERTIES OF GROUTS

In order that a grout may bring about the desired effect it is necessary thattttt it should have the correct fluid properties for injection into the formation, that its set properties satisfy the design specifications, and that the transformation between the fluid and final set states should be sufficiently rapid for displacement of the grout to be unlikely under the stresses to which it will probably be subjected. An additional set property in the case of hydrogels is volume stability against syneresis. Other factors which are important in particular circumstances include the chemical stability of the grout components and their deterioration under exceptionally high or low site temperatures.

A full description of a grout being evaluated for treatment of a particular soil includes its initial density, fluid viscosity and shear strength, particle size distribution of any solids, and any changes in these that may occur during injection. The possibility of the grout mixing with or being diluted by the groundwater, as well as the effect of the fluid on the chemistry of the soil and groundwater, also need to be known.

Viscosity and rigidity are the two rheological properties which govern the flow of grout in the voids of the soil. In suspension grouts they are inversely proportional to the water/solids ratio (Figure 9.2(a)). Any increase in the viscosity of cement grouts with time is sufficiently slow for it to have no effect on injection. The viscosity of chemical grouts varies with the concentration of the reactive chemicals (Figure 9.2(b)) and as some of these grouts contain minute particles in suspension it is perhaps more correct to use the term *apparent viscosity* in such cases. Other factors apart (e.g. dilution by groundwater or reaction with components it carries), gel time depends on the concentration of activator, inhibitor and catalyst in the grout formulation. With most chemical grouts, the gel time can be changed by varying the concentration of one or more of these three components. Increases in temperature can reduce initial viscosity but the reductions are marginal and are quickly compensated by the accelerated gelling process.

Grout which has been injected successfully must develop sufficient rigidity to remain in place. In particular, it must be capable of resisting the hydraulic pressure exerted by groundwater that has been sealed off by the grouting operation. Rigidity of an ordinary clay grout cannot be

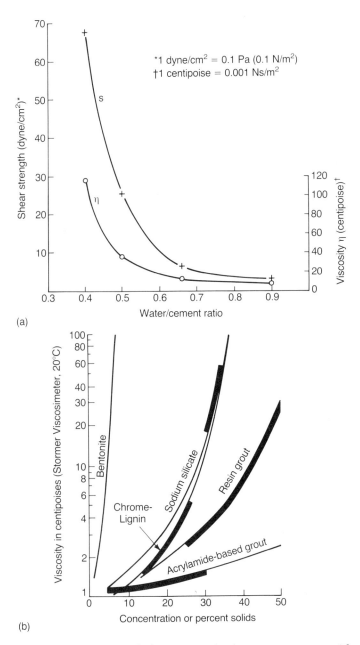

Figure 9.2 (a) Viscosities and shear strengths for cement pastes with varying water/cement ratios (after Raffle and Greenwood, 1961). (b) Viscosities of various grouts; heavy lines indicate the solution concentrations normally used in the field (after Karol, 1982, with permission of ASCE)

improved by adding more of the basic clay constituents as viscosity would increase. This is usually done by adding bentonite (1–10 g/litre) or by using sodium silicate with a reagent. Sodium silicate tends to thicken the grout and increases rigidity markedly.

The yield point of a grout is the lowest stress that must be applied to initiate laminar flow. Consequently, it governs the minimum pumping pressure required to inject grout. The yield point of many suspensions increases if they remain at rest and diminishes when they are agitated. Such suspensions are described as being *thixotropic*. Thixotropic behaviour is most notable in the case of bentonite grouts. One of the problems with thixotropic clay suspensions is that there is no control over the rate of gelation. Hence the fluid properties of such grouts generally are modified and their shear strength reduced during injection by the addition of a dispersing chemical. In this way injection pressures are not dissipated by the need to overcome yield values but only in overcoming viscosity. This enhances the rate of flow.

The important properties of fresh particulate grouts are consistency, fluidity, water retentivity and bleeding. Consistency is a function of resistance to shearing force and is related to the viscosity coefficient. The fluidity is inversely proportional to grout density and determines the velocity of flow under given conditions and given pressure. The more fluid a grout is, the less the pressure loss when moving so that the grout will travel further from the point of injection. Generally fluidity is reduced or enhanced according to the amount of water present. In particulate grouts the resistance to flow is related to the specific surface of the suspended materials or their shape. High specific surfaces, in the case of cementitious grouts, impart greater chemical reactivity. Particle shape can affect flow resistance since spherical shapes tend to reduce friction by imparting a lubricating type of action. There are two other factors that influence flow: namely, interparticle attraction or flocculation, which increases flow resistance, and chemical reactivity of the suspended material in water which, by gelling or viscosity alteration, also increases flow resistance.

The flow properties of a cement grout are affected primarily by dynamic interparticle forces of attraction and repulsion, and in dense grouts by dilatancy of the moving particles. A dense grout can only be pumped easily when it contains sufficient fluid to prevent expansion of the particle matrix during shear. Generally a well-graded range of particles is preferred since the better the grading, the lower the critical porosity at which the grout becomes pumpable. A reasonable percentage of fine particles is also desirable to increase the specific surface of the grout particles and thereby slow the separation of solid and liquid phases.

The movement of fluids through the voids in a soil is resisted by drag at the interface between the grains and the fluid. For true (Newtonian) fluids the drag is proportional to viscosity and shear rate (the shear rate being determined from mean flow velocity and the geometrical characteristics of the void space). Newtonian fluids possess no shear strength. When a grout which possesses shear strength is pumped into the ground under a constant pressure, the opposing drag, due to corresponding shear stress acting at the growing area of the surface wetted by the grout, eventually equals the applied pressure so that none is available to maintain flow. In such instances the initial rate of penetration is determined by the velocity of the grout.

Water retentivity of a grout is indicated by its ability to retain water against vacuum filtration. In cement grouts, the lower the water/cement ratio, the greater is the retention of water. The use of water-reducing agents improves water retentivity.

A particulate grout must be stable, that is, it must be able to remain homogeneous for a predetermined amount of time. Lack of stability leads to separation of the constituents, the particles compacting to form a filter and so giving difficulties during injection. Stability, therefore, is a property of suspensions, other than thixotropic suspensions, which enables them to preserve their original rheological characteristics. It depends on the size and specific surface of the particles along with the intensity of surface action between the particles and the liquid. In the case of cement grout, the addition of a very small percentage of bentonite considerably improves its stability. Stability may be improved by the use of dispersing agents, which also affect fluidity. As far as clay grouts are concerned, their stability also depends on their thixotropy and rigidity. Electrolytes, such as potassium nitrate, potassium carbonate, sodium aluminate, sodium silicate and sodium hydroxide, are commonly used to stabilize (peptize) clay grouts. Bentonite, in very small concentrations, is also used.

In the case of cement grouts, during mixing the particles are dispersed and suspended in water. Except in the case of a very dense paste, the resultant suspension is initially unstable and the cement particles settle under gravity. Bleeding refers to particles coming out of suspension before the grout has cured. The bleeding mechanism is important in grout design, particularly the bleeding rate and the bleeding capacity (i.e. the final volume of bleeding; Figure 9.3). The greater the degree of dispersion, the smaller is the volume of segregation. If bleeding water accumulates under individual particles of a cement grout it destroys their ability to bond and so reduces both the strength and impermeability of the grouted mass (King and Bush, 1958). Hence low bleeding, together with active expansion while setting, are important properties.

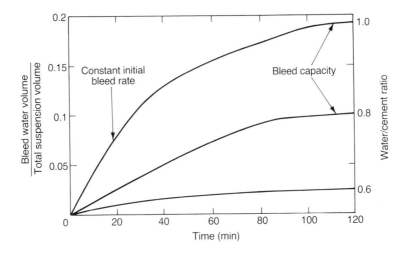

Figure 9.3 Rates of bleed for ordinary Portland cement (after Littlejohn, 1982, with permission of ASCE)

Expansion of a cement grout, can be brought about by the formation within it of minute bubbles of hydrogen. These are generated by the addition of aluminium powder which reacts with the alkalis in the cement. The amount of aluminium powder required is around 0.01% by weight of the cement (i.e. a few grams). Expansion improves the bond between the grout and the interstices into which it is injected.

Cement grouts have well-defined shear strengths that develop immediately after mixing. The shear strength affects penetrability and Table 9.1 gives values of limiting penetration for soil permeability of 10^{-3}, 10^{-4} and 10^{-5} m/s. The shear strength, however, is not a limiting factor inhibiting penetration in coarse-grained soils such as very open gravel.

Table 9.1 Limiting penetration of shear strengths of cement grouts (after Raffle and Greenwood, 1961)

Shear strength (dyne/cm²)	Limiting penetration for 30 m of injection head (m)			Corresponding water/cement ratio for ordinary Portland cement
	$k = 10^{-3}$ m/s	$k = 10^{-4}$ m/s	$k = 10^{-5}$ m/s	
6.6	4.3	1.4	0.5	0.4
25.6		3.6	1.2	0.5
67.6			4.4	0.66

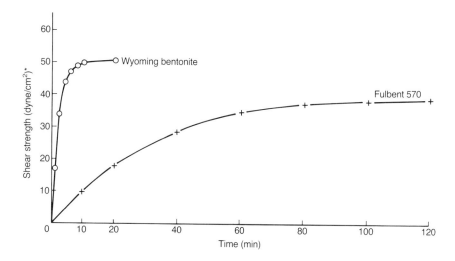

Figure 9.4 Development of shear strength in 5% aqueous solutions of two clays after mixing (after Raffle and Greenwood, 1961)

Clay grouts also exhibit shear strength (Figure 9.4) but violent shearing action temporarily destroys the bulk shear strength. The extent to which it subsequently redevelops is governed by both the level of subsequent rate of shear and the time for which the subsequent rate of shear has been maintained.

Cement grouts do not have a controllable gelling time, whereas true fluid grouts normally do. In some suspensions the gelling action is speeded up by agitation. This is termed *rheopexy* and can cause serious trouble in mixing and injection. Adequate use of dispersants can remedy minor occurrences, but materials exhibiting a high degree of rheopexy should be abandoned.

A clay grout develops a gel after a given interval of time. Nonetheless, even bentonite grout does not develop a stable gel without the assistance of additivies such as cement or silicates. The gel is often thixotropic.

The gel time or induction period of a chemical grout is defined as the time between the initial mixing of the chemical components and the formation of a gel. The choice of induction period for a specific application depends upon many factors. Often the two controlling factors are the volume of the grout to be injected and the pumping rate at the allowable pumping pressure. The permeability of the ground, groundwater conditions and temperature are also important factors. The gel time of most chemical grouts can be varied from minutes to hours but is temperature sensitive. In normal circumstances gel times of 45–90 min are used to give adequate time for mixing, pumping and placement. Where setting times are less than 30 min (or ambient temperatures are high, i.e. greater than 30°C, as setting time is temperature sensitive), proportioning pump systems may be used which delay mixing of chemical components until the point of injection. Gel times for very coarse soils should be relatively short, possibly not exceeding 15 or 30 minutes. On the other hand, in fine sands and silts, gel times of several hours can be used without seriously affecting the efficiency of stabilization.

Some chemical grouts, after catalysation, maintain a constant viscosity and at the end of the induction period turn from liquid to gel almost instantaneously. Others, such as silicate grouts, maintain their initial viscosity for less than the induction period and yet others increase in viscosity from the time of catalysis to gel formation. In the case of these latter two groups of grouts it is not possible to inject the grout into the soil during the whole of the induction period.

Most applications of grout are into moist ground or below the water table so that the grouted mass never dries out. Hence the most significant strength factor is the 'wet' strength. However, dry grouted soils possess higher strength than saturated soils, often by a factor of 10. Apart from the type of grout used, the strength of a fully grouted soil depends on its density, average grain size and grain size distribution in that strength increases with increasing density and decreasing effective grain size (D_{10}). Well-graded soils have higher strengths when grouted than uniform soils with the same effective grain size. The strength also depends upon the extent to which the pores in the soil have been filled.

From Table 9.2 it can be seen that the minimum shear strength for a grout in soil of a given permeability is related to the applied hydraulic gradient. The demand for a high shear strength in a set grout is greatest in soils of high permeability. When the hydraulic gradient is very high the choice of grout must take into account the creep properties of the set grout.

The development of the shear strength of a cement grout is affected by the water/cement ratio (Figures 9.2(a) and 9.5) and the particle size – the larger the water content and particle size, the lower the strength. The rate of shear strength development can be changed by the addition of accelerators or developers. Under optimum curing conditions, and when not subjected to chemical attack, a set grout continues to increase in strength over a prolonged period. Cements with a low rate of hardening tend to have a higher ultimate strength due to slow formation of denser gel during the initial stages of setting. Generally, ordinary Portland cement grout has a set strength at 28 days of approximately 60–70% of ultimate strength (Figure 9.5). This amount of strength may be obtained in about three days with high early strength cement.

Table 9.2 Hydraulic gradient to maintain flow in non-Newtonian grouts (from Bell, 1975)

Soil permeability (m/s)	Yield value (N/m^2)	Hydraulic gradient
10^{-2}	1	1.2
	10	12
	100	120
	1000	1200
10^{-3}	1	4
	10	40
	100	400
	1000	–
10^{-4}	1	12
	10	120
	100	1200
	1000	–
10^{-5}	1	40
	10	400
	100	4000
	1000	–

Chemical grouting of granular soils increases their cohesion but the angle of internal friction remains more or less the same (Figure 9.6). Nevertheless, most chemical grouts when set form weak solids. On the other hand, some special-purpose grouts which contain high concentrations of polymerizing components form very rigid products when set, although at the expense of fluidity during injection. For instance, crushing strengths typical of strong concretes can be produced by injection of sands with grouts based on epoxides or polyesters. If weak grouts are injected into large pores, and a high hydraulic gradient exists, then they may be extruded bodily.

Chemically grouted soils are likely to be subject to creep. According to Karol (1982) it is possible to define a creep endurance limit below which failure does not occur, regardless of the duration of the load. This limit is approximately 25% of the unconfined compressive strength of the materials concerned in situations where faces are exposed, as in open excavations, tunnels and shafts. Where lateral support is afforded to the grouted soil the creep endurance limit approaches half the unconfined compressive strength.

Shrinkage of cement grout is related mainly to the amount of water removed. However, shrinkage is not normally a serious problem since the environment into which grout is injected is often damp or beneath

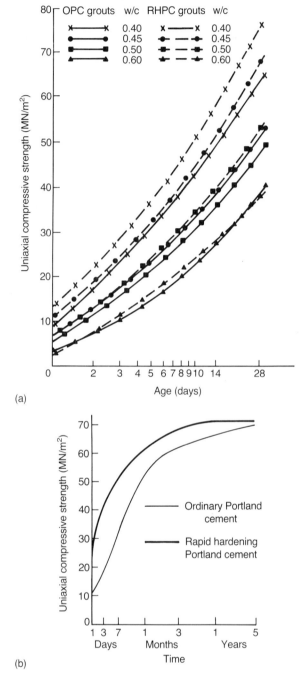

Figure 9.5 (a) Effect of water/cement ratio on strength development of ordinary Portland cement (OPC) and rapid hardening Portland cement (RHPC). (b) Gain in strength of set grouts. (From Littlejohn, 1982, with permission of ASCE)

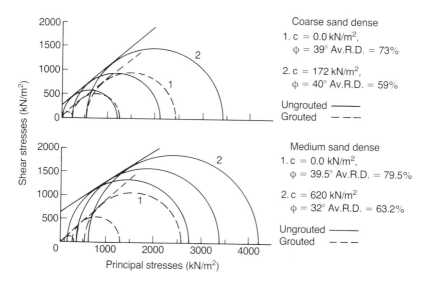

Figure 9.6 Drained triaxial test results for silicate grouted coarse and medium sands (after Skipp and Renner, 1963)

the water table, and moist-cured grout which remains moist does not shrink, in fact it may expand slightly with time.

Cement grouts are durable under most normal conditions but deterioration may be caused by abnormal environmental conditions, such as an attack by sulphates contained in groundwater or prolonged exposure to salt water. Durability may be increased by use of blast furnace or sulphate-resisting cement.

Many single-fluid chemical grouts are subject to syneresis, that is, the expulsion of water from the gel. If this occurs to a significant extent, then seepage channels develop within the grout itself. In silicate grouts in which syneresis occurs, water exudes from the gel within a few hours of setting and the process stabilizes within 3–4 weeks. Syneresis can be controlled in the mix by increasing the concentration of silicate and the coefficient of neutralization. In practice the amount of syneresis associated with a given grout depends upon the ratio of its volume to surface area of the grouted soil since bonding of the gel and solid surface resists internal shrinkage stresses and so reduces the change in grout volume. Hence, grouts that may be unsuitable for coarse gravels may be appropriate for fine sands.

All grouts that contain water not chemically bound to the grout particles are subject to mechanical deterioration if subjected to alternative freeze–thaw and/or wet–dry cycles. The rate at which deterioration occurs varies with the amount of free water available in the grout, as well as the degree of drying or freezing. Deterioration of chemical grouts can occur if the grout reacts with the soil or groundwater (Karol, 1982). Chemical grouts based upon aqueous solutions may dissolve on long-term contact with groundwater, dissolution governing the permanence of the grout treatment. The soundest defence against dissolution is to ensure maximum void filling which precludes access of aggressive groundwater to grout.

In summary, the choice of grout(s) for a particular project depends on its basic properties. In the case of particulate grouts, they must be stable during mixing and injection. Ideally the setting time should be controllable to allow the grout to set in the right place at the right time. The particle size imposes a lower limit to the size of void it can penetrate. The viscosity of the grout provides a measure of the degree to which a grout can penetrate fine pores and its fluidity facilitates pumping. For grouting by permeation, the lowest viscosity grout that yields adequate strength tends to be selected for use in formations of moderate to low permeability. The other flow properties and the gel time determine the maximum value of the radius of injection. The importance of the gel strength depends on whether the formation requires strengthening or sealing. The gel strength should be high enough to resist any tendency to creep. The grout when set must resist chemical attack and erosion, and displacement by groundwater.

9.3 PENETRATION OF GROUTS

The ability of grout to penetrate soil depends on the particle size distribution, permeability and porosity of the soil; the pressure being used for injection; the period of injection; and the viscosity of the grout. Because grouts do not maintain the same viscosity indefinitely, there is a limit to the distance any grout can penetrate before the decreasing head available at the advancing interface between the grout and the pore water, and the increasing viscosity of the grout as it sets, prevent further flow.

The penetration characteristics of cement or clay grouts or admixtures thereof are very different from those of chemical grouts. When particulate grouts are injected into porous soil, filtering may occur whereby the larger particles in the suspension tend to separate out at the entrances to pores. For example, filtercake formation occurs when cement grout is injected into uniform sands of less than 2 mm grain size.

Hence the ability of particulate grouts to penetrate a formation depends upon the particle size of the suspended material in that the particles of the grout must be smaller than the voids that require filling. This has been indicated in terms of a groutability ratio, which has been defined as

$$N = \frac{D_{15} \text{ (soil)}}{D_{85} \text{ (grout)}} \tag{9.1}$$

This ratio should exceed 25 if a grout is successfully to penetrate the formation concerned. Grouting is not possible if the ratio is less than 11. However, Burwell (1958) suggested that the D_{85} value of the grout should not be relied upon solely. He recommended that when the limits of groutability are approached, the criterion should be re-examined on the basis of

$$N = \frac{D_{10} \text{ (soil)}}{D_{95} \text{ (grout)}} \tag{9.2}$$

in order to be doubly sure that the grout is suitable. In the latter case grouting is consistently possible when N values are above 11 but is impossible when they are less than 5. Alternatively, the limits for particulate grouts may be taken as a 10:1 size factor between the D_{15} of the grout and the D_{15} size of the granular soil concerned.

Soils containing less than 10% fines can usually be permeated with chemical grouts. If the fines content exceeds 15%, effective chemical grouting may prove difficult. Permeation grouting is not possible when the fines content is greater than 20%. Then hydrofracture must be used.

Particle fineness (given in cm²/g) also offers an indication of groutability. For example, Nonveiller (1989) quoted limits of penetration for ordinary Portland cement (OPC); high early strength cement (HESC); colloidal fine cement (CFC); and ultra-fine cement (UFC) into granular soils. These are given in Table 9.3.

Alternatively, permeability may be used to assess groutability. For example, Littlejohn (1985) suggested that cement grouts cannot be used for treating soils with permeabilities less than 5×10^{-3} m/s as cement particles would be filtered out during injection. As far as clay grouts are concerned, the small sizes of the particles permits them to penetrate voids in soil with a permeability of 10^{-5} m/s. The limits of groutability of some grout mixes are given in Table 9.4. Initial estimates of the

Table 9.3 Limits of penetration of cement grout into granular soils

Type of cement	Specific surface of cement (cm²/g)	Permeability, k (m/s)	D_{85} of cement (mm)	D_{15} of soil (mm)
OPC	3.17	2.3×10^{-3}	0.047	0.87
HESC	4.32	1.3×10^{-3}	0.333	0.67
CFC	6.27	3.2×10^{-4}	0.019	0.38
UFC	8.15	3.5×10^{-5}	0.006	0.12

Table 9.4 Limits of groutability of some grout mixes (after Caron *et al.*, 1975)

	Types of soils		
Characteristics	*Coarse sands and gravels*	*Medium to fine sands*	*Silty or clayey sands, silts*
Grain diameter	$d_{10} > 0.5$ mm	$0.02 < d_{10} < 0.5$ mm	$d_{10} < 0.02$ mm
Specific surface	$S < 100$ cm	100 cm $< S < 1000$ cm	$S > 1000$ cm
Permeability	$k > 10^{-3}$ m/s	$10^{-3} > k > 10^{-5}$ m/s	$k < 10^{-5}$ m/s
Type of mix	Bingham suspensions	Colloid solutions (gels)	Pure solutions (resins)
Consolidation grouting	Cement $(k > 10^{-2}$ m/s) Aerated mix	Hard silica gels: double shot: Joosten (for $k > 10^{-4}$ m/s) ——— single shot: Carongel Glyoxol Siroc	Aminoplastic phenoplastic
Impermeability grouting	Aerated mix Bentonite gel Clay gel Clay–cement	Bentonite gel Ligno-chromate Light carongel Soft silicagel Vulcanizable oils Others (Terranier)	Acrylamids Aminoplastic Phenoplastic

groutability of ground have frequently been based on the results of Lefranc tests.

The penetrability of chemical grouts is governed by their viscosity, injection pressure and period of injection, as well as the permeability of the soil being grouted (Bodocsi and Bourers, 1991). Chemical grouts with viscosities less than 2×10^{-3} N s/m^2 (2 cP) (e.g. acrylamide-based grouts) can usually be injected without difficulty into soils with permeabilities as low as 10^{-6} m/s. At 5×10^{-3} N s/m^2 (5 cP) grouts (e.g. chrome–lignin grouts) are limited to soils with permeabilities higher than 10^{-5} m/s. At 1×10^{-2} N s/m^2 (10 cP) grouts (e.g. silicate-based formulations) may not penetrate soils below 10^{-4} m/s. For higher viscosities, for example, 2×10^{-2} N s/m^2 (20 cP) it may be necessary to restrict the grout application to more permeable ground or reduce the hole spacing.

In the case of particulate grouts, penetration is also governed by their shear strength in that their initial shear strength must be overcome before the grout begins to flow. The critical shear stress needed to initiate flow of cement and clay grouts falls within the range 1–20 N/ m^2. Although this represents an extremely weak material, in permeation flow the modest shear stresses involved must be summed over the very considerable surface area of the passages. Once started, flow is directly proportional to the excess shear stress.

9.4 TYPES OF GROUT

9.4.1 Cement grouts

Cement grouts may consist of water and cement only, or water, cement and other materials that combine chemically with the cement for special purposes or that serve as bulking agents. The most widely used form of grout is a mixture of ordinary Portland cement and water. Its popularity is due to its ready availability, its reproducible performance, its high yield strengths, and it is far cheaper than any chemical grout (Anon., 1990(c)). Its disadvantages are slow gain of strength, and particle size which precludes permeation into soils of permeability less than 10^{-3} m/s.

The principal factor affecting the properties of cement grouts is the water/cement ratio, the amount of water governing the rate of bleeding, subsequent plasticity and ultimate strength of the grout (Littlejohn, 1982). The extent to which these and fluidity are related to the water/ cement ratio of neat cement grout is shown in Figure 9.7. Cement grouts tend to undergo excessive bleeding at water/cement ratios greater than 1.0. They also have low strength, increased shrinkage and poor dura- bility. Obviously grouts with high ratios are more difficult to inject but they undergo less segregation and give higher strength than those with lower ratios.

Cement grouts have low viscosities and no rigidity for water/cement ratios higher than 0.5 (by weight). The viscosity of the grout increases sharply and it acquires some rigidity with lower values of water/cement ratio. For a given cement, its density after setting is fairly constant and is independent of the water/cement ratio of the grout. The den- sity varies according to the type of cement used. Nevertheless, the volume of hardened cement grouts that have identical proportions and are mixed under similar conditions may vary by as much as 50%. The setting time of cement grouts increases as the water/cement ratio in- creases. For example, cement grouts tend to set after about 4–5 hours, but when they have been greatly diluted their setting time may amount to 10–15 hours. Some cements never set for water/cement ratios higher than 10.

Water/cement ratios varying from 0.5:1 up to 10:1 have been used (Anon. 1962). For instance, if the soil contains large voids which accept

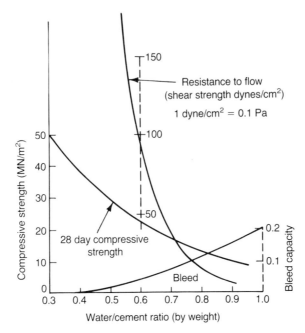

Figure 9.7 Effect of water content on grout properties (after Littlejohn, 1982, with permission of ASCE)

grout readily, the water/cement ratio may be 0.5:1. The usual water/cement ratios fall within the range 0.8:1 to 5:1.

Pure cement grouts are unstable. As a consequence their flow velocity diminishes rapidly with increasing distance from the grout-hole and particles settle out at a rate that decreases with the water/cement ratio of the grout. Hence, as a general rule the initially injected grout should be fairly thin (water/cement ratio of 10:1 to 15:1) in order to observe the behaviour of the grout-hole and the ground, and to minimize the chances of prematurely plugging the hole by too thick a grout. If the grout-take becomes excessive without building pressure, the hole should be allowed to rest a sufficient time to allow initial set before continuing.

In a pure cement grout, the particles of cement are held in suspension only by the agitation of the mixer or the turbulent flow through the pipes which, for this purpose, must be of small diameter, generally 25 mm (Cambefort, 1977). Since the diameter of the grout-hole is larger, the particles of cement gradually block it up, leaving channels in which the flow remains turbulent. But these channels may become blocked before the ground is properly injected. In such instances the refusal pressure which is suddenly attained is artificial. In order to resume injection it is necessary to redrill the grout-hole.

Table 9.5 Common additives to cement grout used to impart specific properties to the final product (after Littlejohn, 1982)

Admixture	Chemical	Optimum dosage (% cement wt)	Remarks
Accelerator	Calcium chloride	1–2	Accelerates set and hardening
	Sodium silicate	0.5–3	Accelerates set
	Sodium aluminate	0.5–3	Accelerates set
Retarder	Calcium ligno-sulphate	0.2–0.5	Also increases fluidity
	Tartaric acid	0.1–0.5	
	Sugar	0.01–0.5	
Fluidizer	Calcium ligno-sulphonate	0.2–0.3	
	Detergent	0.05	Entrains air
Air entrainer	Vinsol resin	0.1–0.2	Up to 10% of air entrained
Expander	Aluminum powder	0.005–0.02	Up to 15% preset expansion
	Saturated brine	30–60	Up to 1% postset expansion
Antibleed	Cellulose ether	0.2–0.3 (for $w < 0.7$)	Equivalent to 0.5% of mixing water
	Aluminum sulphate	Up to 20% (for $w < 5$)	Entrains air

Cement grout can be improved by mechanical or physico-chemical action. High-speed mixing (1500–3000 rev/min) activates the hydration of cement particles, gives a better dispersion and brings about a rapid formation of small crystalline elements of hydrates of different types. Grouts obtained through energetic mixing are sometimes called *activated* or *colloidal* grouts. Basically these grouts are more stable and more fluid than ordinary grouts, but only for low water contents.

Other materials may be added to cement grout to control, to a limited degree, the physical properties of the mix, such as its shrinkage, its strength and its setting time (Klein and Polivka, 1958). Basically these admixtures (Table 9.5) are as follows:

1. Accelerators for hastening the setting time.
2. Retarders for delaying the setting time.
3. Lubricants for increasing flowability.
4. Protective colloids for minimizing segregation.
5. Expansion materials for minimizing shrinkage.
6. Water-reducing materials for reducing the water/cement ratio.

Admixtures for grouts include fillers such as sands, clays and pozzolans. The main purpose of fillers is to reduce the cost of the grout. Certain fillers, however, offer certain advantages such as reduced bleeding, or improved fluidity or retardation. Fine sands can be added to cement grouts when large voids have to be filled.

Grouts with a very wide range of strengths can be produced by using clay in combination with cement. Bentonite–cement grouts are stable over almost any water/cement ratio, provided that a suitable bentonite content is used. The dispersed bentonite in water acts as a suspending agent preventing cement particles settling out. Usually about 3% of bentonite causes a marked increase in viscosity on standing and the tendency is to water down the mix to the consistency that would be expected with a neat cement grout. Addition of cement to a thixotropic bentonite–water mix gives it a consistency that varies from soft plastic to hard rigid strong mortar, depending on the water/cement ratio. Bentonite holds cement in suspension while hydration takes place, so that the whole grout sets to form a hard cementitious solid. Owing to its fineness bentonite has a lubricating quality which allows suspensions of low water/cement ratio to be pumped, so that little or no surplus water is given off, thus allowing voids to be filled completely in a single operation.

Pozzolans, notably pulverized fly ash (PFA), when combined with ordinary Portland cement contribute to the cementitious properties of the hardened grout. Pulverized fly ash has a similar particle size range to ordinary Portland cement. The addition of such rounded particles to cement improves the flow properties, thus helping penetration. The PFA/cement ratios in common use vary from 1:4 to 20:1 depending upon the strength and elastic properties required.

Water-soluble compounds are chemical reagents that are employed in relatively minute quantities from 0.01 to 2% by weight of ordinary Portland cement. They include surface-active agents (water-reducing, air-entraining and dispersing agents), retarders and accelerators. Such agents may be employed either alone or in combination to enhance the properties of fresh and hardened grouts.

Surface-active agents improve the fluidity of a grout, which means that the water/cement ratio may be reduced significantly yet the grout maintains the same consistency. Such additives can facilitate the penetration of grout. Moreover, because they lower the water content they decrease the likelihood of bleeding and improve strength and watertightness. Water-reducing agents include calcium, sodium or ammonium salts of ligno-sulphates. At higher water/cement ratios surface-active agents may aid dispersion, whereby particles in the grout are kept separate. Many surface-active reducing agents (e.g. lignins) retard hydration of cement, which may be desirable in warm weather.

Accelerators, which may be used in cold weather to hasten setting, include calcium chloride and alkali carbonates and hydroxides.

When colloidal compounds are added to cement grouts they act as thickeners and, as such, increase the water requirement of the grout at a given consistency. They reduce bleeding and segregation significantly in thin grouts.

9.4.2 Cement–clay grouts

Cement–clay grouts are most suitable where the main purpose of grouting is to arrest water movement (Johnson, 1958). When grouting gravels, some sand or silt may be included in the mix. Because of the size of cement particles such grouts are used to treat soils with permeabilities in excess of 10^{-3} m/s, such as coarse alluvium. Fluidity is not usually important since the grouts are only used to treat soils sufficiently open to accept the coarser cement particles without filtration. The clay is present in the grout in the role of a filler. The development of strength is slow and there is no well-defined setting time. The set of cement–clay grout can be accelerated by adding either calcium chloride or sodium silicate and reagent. The efficiency of these additives is limited when used with cement–clay grouts having high clay contents, and in such cases the latter accelerator is used. Set grouts have low crushing strengths in relation to those of neat cement grouts. Actual strengths range from less than 7 kN/m^2 to over 7 MN/m^2, increasing with cement content. The ability of clays to form gels in grout helps stabilize the cement, significantly slowing down its settlement from suspension and its bleeding from the grout. The stability of cement–clay grouts is directly proportional to the quality of the clay and its proportion in the grout. A small amount of clay is usually required to obtain grouts which settle with little or no water gain. Grouts with no water gain are keenly sought after since they usually fully occupy the voids during the first grouting operation so that no redrilling and regrouting is required. Grouts which settle with no water gain sometimes can be thixotropic, that is, until the cement hardens. Cement–clay grouts always have some rigidity. Cement–clay grouts have relatively high yield values (100–500 dyne/cm^2; 10–50 N/m^2) which require correspondingly high injection pressures.

9.4.3 Clay grouts

Natural clays are used whenever possible for clay grouts for reasons of economy. The particle size distribution of a clay used for grouting is important. Normally a clay with a liquid limit of less than 60% is not considered unless the coarser fraction can be removed economically.

The relevant rheological properties of a suitable clay for grouting are more or less determined by the colloidal content. The higher the content of clay minerals in a deposit of clay, the better it is for grouting purposes. Montmorillonite is the most active clay mineral and has the capacity to swell significantly. Viscosity, rigidity and thixotropy can be affected by adding either electrolytes, colloidal material (bentonite) or wetting and dispersing agents.

Clay suspensions by themselves are usually slightly alkaline and tend to flocculate on addition of acidic materials. The pH value of the grout should exceed 7, if not it should be adjusted by the addition of dispersants such as sodium phoshate (Kravetz, 1958). Clay grouts with added dispersing chemicals almost invariably incorporate a third ingredient so that shear strength is restored as chemical reaction proceeds and the grout can develop a permanent set.

Clay grouts do not have a high strength but reduce the permeability of the ground and offer high resistance to displacement by water gradients. With neat clay grouts, particularly bentonite grouts, the shear stress (yield value) that has to be applied during injection to overcome thixotropic gelling and transform the suspension to a fluid leads to an unwelcome increase in injection pressures. Use of a dispersing agent reduces the viscosity, yield stress and gel strength, and makes the grout more easy to inject. Inclusion of a setting agent can control the set time within certain limits. The setting agent is normally a sodium silicate.

Bentonite is a moisture-absorbing, colloidal clay which, because of its small particle size (less than 2 μm), has been used in grouting coarse to medium sands (Ischy and Glossop, 1962). The gel strength of these grouts is not sufficient to give appreciable increase in strength to the soil. Hence, they generally are used only for sealing or reducing permeability. However, the injection of bentonite grout into sands may be complicated by thixotropic effects, and localized gelation in small pores can cause flow anomalies. Bentonite is easily pumped when mixed with water and it can absorb three or more times its dry bulk volume.

At rest bentonite suspension undergo thixotropic gelation. If stirred vigorously in a mixer, the suspension flows like a mobile liquid, the viscosity decreasing as the rate of stirring increases. When the motion ceases the suspension begins to set to a gel which stiffens progressively. Gelation is reversible and the sol-gel transformation may be repeated indefinitely. The rate of gelation, but not the final set strength, is influenced markedly by temperature. The gel shrinks irreversibly on dehydration and on freezing. The strength of a thixotropic gel depends on the bentonite concentration, the setting time and the chemical composition of the suspending fluid. The shear strength of a thixotropic suspension varies with time and may be much higher than its yield value.

Flocculation may be offset by the addition of dispersing agents, thereby providing a means of controlling the yield value. In addition, dispersing agents (for example, sodium polyphosphate) extend the range of composition over which controllable setting can be obtained and may act as retarders.

The addition of sodium silicate to bentonite suspensions leads to a reduction in yield point and thixotropy, when the clay and silicate concentrations in the suspension are low. At higher silicate concentrations progressive gelation of the suspension occurs on standing. The gel formed in this way is permanent and non-thixotropic. The rate of gelation and the final shear strength of the gel are markedly influenced by the concentration of bentonite and sodium silicate in the suspension, and by the shear history of the suspension before it begins to set. Gels of appreciable mechanical strength can be produced, with setting times of 0.1–5 hours. No shrinkage occurs during the setting process.

9.4.4 Clay–chemical grouts

Clay–chemical grouts contain clay as the major component. The most common chemicals used in these grouts are sodium silicate with either sodium aluminate or hydrochloric acid as reagents (King and Bush, 1963). The chrome–lignins are also satisfactory gel-forming compounds and are available either as crude sulphite liquor and chromic acid in liquid form or as a preblended powder that only needs to be mixed with water. Phosphates may be used to facilitate mixing of bentonite powder to ensure its full dispersal in reasonable time.

Clay–chemical grouts are stable and possess low yield values and viscosities which enable them to be injected into fine- to medium-grained sands to reduce permeability. In fact these grouts represent the lower limit of the suspension grouts that can be injected into sands. Their ability to penetrate the ground depends upon initial viscosity and size of clay particles. Bentonite is the finest of the clays, superbentonite possessing even finer particle grading. Sodium bentonites react best with chemical additives.

9.4.5 Chemical grouts

Chemical grouts were defined by Anon. (1957) as true solutions which contain no suspended solid particles unless deliberately added for some specific purpose. Hence, they are able to penetrate fine sands and sandy silts. However, they are expensive. They can be used as the final seal after a preliminary stage grouting has been completed with a cheaper grout. Chemical grouts are Newtonian fluids and, as such, lack shear strength. They are available in a wide range of viscosities ranging down to very thin liquids.

Information on the size of openings, the coatings on the surface of openings and the amount of free water or moisture present must be available before the type of chemical grout can be chosen. For instance, passages as narrow as 0.05 mm have been grouted and some chemical grouts bond poorly to wet or even moist surfaces. The pot life (time between mixing of components and the start of set) of different chemical grouts varies widely from 5 min to several hours. This must be considered when the grout and method of injection are selected for a specific project. Furthermore, the temperature of the air and ground must be known, as they affect the pot life and viscosity of the grout and, hence, the ability to penetrate voids. Wherever feasible, groundwater from the site, at the site temperature, should be used to prepare the stock solutions to eliminate differences in tank and underground gel times.

Chemical grouts can be used in two main ways (Karol, 1982). One is a 'two-shot' process in which two different chemicals are brought into contact to form an insoluble precipitation in the voids. In the two-shot process the more viscous of the two reactants normally is injected first, followed by injection of a second, less viscous, chemical which mixes and reacts with the first to produce the precipitate (for example, the Joosten process uses sodium silicate with calcium chloride as the gelling agent). Two-shot silicate systems have largely given way to one-shot systems. In this process all the ingredients are premixed prior to injection and the grout is so designed that reaction to convert the chemical into a solid or gelatinous mass takes place in the void (Mitchell, 1970). It is believed that for both types of grouts virtually all the in-place strength is derived by mechanical bonding between the treated material and the chemicals.

The two-shot process generally yields much higher unconfined compressive strengths than the one-shot process. For instance, strengths of between 2.8 and 7.0 MN/m^2 are commonly associated with the Joosten process. The gel undergoes almost no shrinkage on setting, it resists temperature changes and saline and sulphate attack, and is non-toxic. On the other hand, the two-shot process suffers the disadvantage of being slower because sodium silicate solution is pumped into the soil as the grout pipe is advanced downwards (Figure 9.8). The pipe then is flushed with water and calcium chloride is pumped in as the pipe is retracted. Because precipitation occurs upon contact between the two solutions this limits the penetrability of silicate grout, necessitating very close spacing of grout pipes. Higher injection pressures are also required to assist easy acceptance of the grout. Perfect interpenetration of the two shots may not always be obtained.

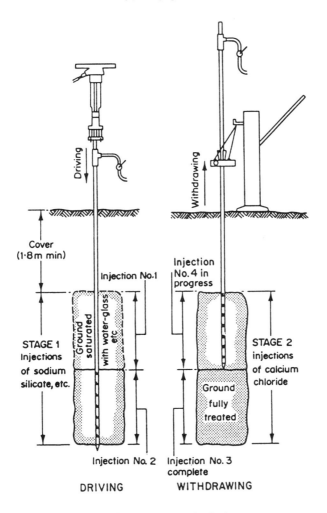

Figure 9.8 The Joosten method of grouting

Generally, one-shot grouts are more versatile than two-shot systems. Their gel times may be reproducibly controlled from a few minutes to several hours, with a wide range of viscosity (1 to 10 × 10^{-3} N s/m^2; 1–10 cP) and penetrability (down to 10^{-6} m/s) can be obtained (Glossop, 1968). The gel strength of one-shot silicate grouts is low (up to 3.5 MN/m^2). However, if organic esters (ethyl acetate) are mixed with sodium silicate, not only is the gelling time delayed but such grout brings about an appreciable increase in strength of the treated soil.

The sodium silicates are the most viscous of the chemical grouts ranging from 2 to 5 × 10^{-2} N s/m^2 (from 20 to 50 cP). On ageing the gel shrinks, becomes opalescent and cracks. Syneresis of normally diluted single-phase silicate gels occurs when they are injected into coarse sands (Figure 9.9). However, Cambefort (1977) reported that when grains are smaller than 1.5–2 mm, the gel retains its properties (e.g. the larger the specific surface of sand grains, the higher the strength of the grouted mass; Figure 9.10). He therefore asserted that

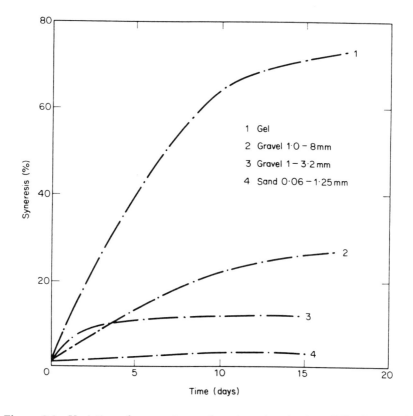

1 Gel
2 Gravel 1·0 – 8 mm
3 Gravel 1 – 3·2 mm
4 Sand 0·06 – 1·25 mm

Figure 9.9 Variation of syneresis as a function of grain size; 60% silicate–ethyl acetate gel (after Caron, 1965)

it is imperative to grout a sand–gravel formation with a clay–cement grout to occupy the large interstices prior to treatment with silicate. Grouts containing 30% silicate are typical for waterproofing applications. Where high strength is required, silicate concentrations of 40–60% are used.

Epoxy and polyester resin grouts are generally supplied as two separate components (base resin and catalyst or hardener) which, when combined with each other, start to react (polymerize) to form a

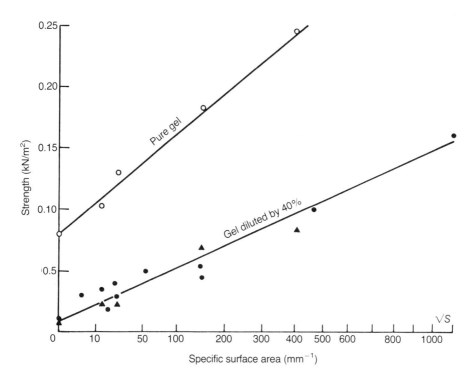

Figure 9.10 Compressive strength of sand–silicate gel mortars. They increase linearly with the square root of the specific surface of sand grains (from Cambefort, 1977)

stiff compound. Various formulations of polyester resins are available so that the properties of the grout may be varied by changing the components. Similarly, the pot life of the grout can be adjusted over a wide range, depending on the temperature, from a few minutes to several hours. Resin grouts have a considerably lower viscosity than silicate grouts and develop somewhat higher strength. The compressive strength of polyester grout ranges up to 140 MN/m^2 and over, with tensile strength of up to 60 MN/m^2. These grouts are capable of developing a strong bond with clean, dry surfaces. However, shrinkage, which can be as high as 6%, occurs during hardening.

Phenoplast resins are polycondensates resulting from the reaction of phenol on an aldehyde. Resorcinol is in this category of resins to be used as grout and it is commonly reacted with formaldehyde (Karol, 1982). A catalyst is required to control the pH, sodium hydroxide normally being used. Setting times vary greatly with the pH of the solution, being shortest when the pH is slightly above 9. The initial viscosity of resorcinol formaldehyde ranges from 1.5 to 3 × 10^{-3} N s/m^2 (from 1.5 to 3 cP) and it remains more or less constant until gelation starts. The change from liquid to gel is almost instantaneous.

Aminoplasts are grouts in which the major ingredients are urea and formaldehyde. Unfortunately these materials will only gel under acid conditions, a distinct disadvantage to their use as grouts. Hence they can only be used when the pH of the groundwater is less than 7.

A range of grouts is based on lignin and sodium dichromate. A chrome–lignin gel is formed by using a dichromate salt to catalyse a calcium ligno-sulphonate solution. Water dilution is used and acceleration is produced by dosage with ferric chloride, which is toxic in the liquid state. A strong elastic gel is produced. Treated soil strengths range from 170 to 700 kN/m^2. Chrome–lignin solutions have viscosities in the 2 to 5 × 10^{-3} N s/m^2 (2–5 cP) range and gel times can be controlled with fair precision in the field. The materials, however, are sensitive to groundwater dilution and acid conditions. In addition, the viscosity of the solution increases gradually from the instant of catalysis until a gel is finally formed.

Acrylamide-based materials come closest in terms of performance to meeting the specifications for an ideal grout. They penetrate more readily, maintain constant velocity during the induction period, have better gel time control and adequate strength for most applications (Karol, 1982). They are, however, more costly than silicate and acrylamide is neurotoxic. Several acrylamide grouts have been withdrawn because of their toxicity (for example, AM-9, Rocagil BT and Nitto SS). The gel, however, is non-toxic.

9.4.6 Bitumen emulsion grouts

Bitumen emulsions can be made having different viscosities, rate of break, and bitumen type and content. An emulsion is a dispersion of minute droplets of a liquid in another liquid in which it is not miscible. A viscous liquid such as bitumen can be dispersed in a non-viscous liquid, like water, to give a relatively non-viscous fluid. Emulsions of bitumen and water may contain 30% of bitumen. Such dispersions can be injected into soils and then broken, that is, the two components separate. The breakdown of the emulsion in the soil can be brought about either by the addition of an organic ester or by means of a synthetic resin. The time required for the emulsion to breakdown after injection can be regulated, according to Cambefort (1977) by the use of casein as a stabilizer and a mixture of sodium aluminate and ethyl formate as a coagulator. Caustic soda can also be used to extend the injection time and to precipitate and control the reaction in different water conditions. As the bitumen can be dispersed as very fine droplets (around 1–2 μm diameter) and as the weight and viscosity hardly differ from that of water, the emulsion can penetrate into granular soil (that is, materials with dimensions exceeding 0.1 mm may be treated). Bitumen is a Newtonian liquid which always flows when subjected to a pressure gradient. This is a drawback as far as long-term treatment is concerned.

9.5 GROUTING

The design of a successful grouting programme requires the selection of a suitable grout material, and the correct drilling equipment, procedures and grout-hole patterns. It is essential that the pipes and injection ports are in the correct place. It is more important to ensure that the full design soil volume is permeated with grout when the objective is water cut-off than when the objective is to improve mechanical properties.

The early stages of a grouting programme are somewhat experimental. Several holes have to be grouted to provide data relating to the pumping pressures that may be safely used, the depth of hole to grout in one stage, the grout mixes to use, the extent of surface leaks and how much grout the injection holes are likely to accept in a given area. As more data are gathered regarding the geological conditions and their behaviour under treatment, so a more definite programme can be planned.

The grout pattern includes the layout of the holes, the sequence in which each hole is placed and grouted, and the vertical thickness and sequence of grouting the stages for each hole. The layout of holes may follow some geometric pattern or this may be modified in relation to the ground conditions. Hole spacings of about 1.3–2.5 m are typical.

Grout-holes in soils may be formed by rotary-percussion, percussion, or augering methods. Generally, cased holes are required in unconsolidated deposits where the hole extends to 3 m or more. Alternatively, a jetted or driven injection pipe can be used for grouting. The driving technique has the advantage over jetting in that it results in a tighter contact between pipe and soil, thereby minimizing the danger of the grout rising to the surface along the outside of the pipe rather than penetrating the soil.

The use of gel times that are less than the pumping time, termed *fast gel times*, has the advantage of grout location control in flowing groundwater and in stratified soils. It also limits grout-takes in very pervious materials. Pumping time to gel time ratios of 10 are common.

The rate at which grout can be injected into the ground generally increases with an increase in the grouting pressure, but this is limited since excessive pressures cause the ground to fracture and lift. The safe maximum pressure depends on the weight of overburden, the strength of the ground, the *in-situ* stresses, the pore-water pressures and the permissible amount of ground surface movement, if any. However, there is no simple relationship between these factors, and so a common rule-of-thumb for safe maximum grouting pressure is to relate the pressures to the weight of overburden. For example, the pressures used, as measured at the top of the hole, may start at 70 kN/m^2 for the first 3.1 m stage and increase by 70 kN/m^2 in each successive 3.1 m stage, while not exceeding 350 kN/m^2 for the fifth and lower stages. Values two to three times as great may be used with fast gel time systems.

9.5.1 Permeation grouting

Permeation grouting at shallow depths may take place at a single stage from a grout pipe, the grout-hole being sunk to full depth and then grouted upwards. Alternatively, grouting may proceed while the hole is drilled. The hole is extended a short distance using a hollow drill rod, it is then withdrawn this distance and grout is injected from the rod. The hole remains open over the length exposed as the soil will have been stabilized sufficiently by the migration of grout from previous injections. The cycle is then repeated. Lastly, grouting is continued as the rod is finally withdrawn from the hole.

Stage grouting is used when relatively high grouting pressures have to be employed to achieve satisfactory penetration of grout in deep holes or tighter sections of holes. In stage grouting the hole is drilled to a given depth and then grouted. After the grout has set the hole is deepened for the next stage of grouting when the procedure is repeated (Figure 9.11). Stage grouting allows increasing grout pressures to be used for increasing depth of grout-hole and reduces the loss of grout due to leakage at the surface.

The *tube-à-manchette* consists of a steel tube, between 37.5 and 62.5

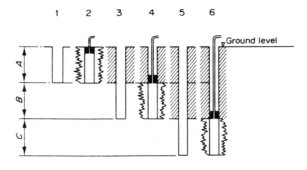

Figure 9.11 Stage-and-packer grouting

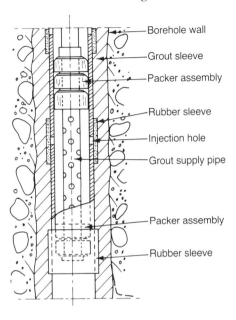

Figure 9.12 Detail of the *tube-à-manchette*.

mm in diameter. This is perforated with rings of small holes (about 8 mm diameter) at intervals of approximately 0.3 m. Each ring of holes is enclosed by a tightly fitting rubber sleeve which acts as a one-way valve (Figure 9.12). A drillhole is sunk, with the aid of casing, to the full depth to be treated and the *tube-à-manchette* placed in it. The casing is withdrawn and grout, termed *sleeve grout* (clay–cement or bentonite), is poured into the annular space left behind. Grouting is then carried out through the *tube-à-manchette* by lowering into it a small-diameter injection pipe perforated at its lower end and fitted with two U-packers. The packers can be centred over any one of the rings of injection holes. When injection starts the pressure in the grout pipe rises until the grout lifts the rubber sleeve, rupturing the sleeve grout and escaping through the small holes into the soil. The rubber sleeves stop any return of the grout into the *tube-à-manchette* and the sleeve grout prevents any leakage of grout at the surface.

Use of the *tube-à-manchette* offers great flexibility since the same hole can be grouted more than once and different grouts can be used. In this way coarser soils can be treated first and finer ones later. Hence the *tube-à-manchette* is often used to grout alluvial soils.

9.5.2 Claquage grouting

Claquage or fracture grouting is frequently used to grout alluvial soils. The aim is to develop a network of grouted fractures as the fine-grained types, silts and clays, are not amenable to penetration grouting (Ischy and Glossop, 1962; Samol and Priebel, 1985). As injection proceeds claquages spread rapidly, forming an intermeshing network of grout-filled fractures and in this way reduce the permeability, as well as compacting and improving the mechanical properties of the soil. Claquages appear when the pressure in the grout-hole exceeds a certain value depending on the characteristics of the soil and depth of overburden. Considerable pressure may be required to start the flow, but once the grout begins to move the pressure can be reduced. In theory this pressure is proportional to the rate of flow per unit length of drillhole, to the viscosity of the grout and inversely proportional to the permeability of the ground. Claquages may form in any direction. According to Cambefort (1977), vertical fractures tend to develop before horizontal fractures. The grout occupying horizontal fractures, in particular, may cause the ground to heave. If damaging heave is to be avoided, ground movement has to be monitored. Claquage grouting can also be used for reducing the amount of settlement (e.g. in fills, see Arcones *et al.*, 1985) or for underpinning to restore the levels of buildings and structures (Raabe and Esters, 1990).

9.5.3 Compaction grouting

Compaction grouting uses highly viscous grout (mixtures of cement, soil, clay and/or PFA, and water) to compress the surrounding soil (Warner and Brown, 1974). The hardened grout forms a bulb or column of strong, relatively incompressible material. Although compaction grouting can be used in any type of soil, it is most frequently used in soils finer than medium-grained sands. One of its advantages is that its maximum effect is obtained in the weakest soil zones. As the size of the grout mass increases this can cause the ground to heave. Thus compaction grouting can be used to correct differential settlements and provide underpinning of structures. It can also be used to strengthen ground adjacent to open excavation or tunnelling operations. To be effective, compaction grouting should not be undertaken at depths less than 1–2 m unless there is an overlying structure to provide confinement.

Grouting may proceed in stages from near the ground surface, downwards to a firm-bearing stratum, or from depth to the ground surface. According to Stilley (1982), if underpinning is the objective, then grouting downwards in stages beneath the building is the more successful. Grouting from the top downwards results in a greater grout-take per hole, and a more complete densification of the lower soils because higher pressures can be used after grouting the overlying soil. Indeed, when a problem soil extends to the ground surface, downward grouting is more or less mandatory in order to prevent grout escaping at the surface. Unless the natural moisture content is substantially on the wet side of optimum, injection of water prior to grouting results in greater grout-takes.

Typical pressures at the point of injection vary from 350 kN/m^2 to 1.7 MN/m^2 when injection is within 2 m of the surface to over 3.5 MN/m^2 when grouting takes place at depths greater than 6 m. Pressures above 4.2 MN/m^2 at the injection point are seldom exceeded irrespective of depth, although pressures as high as 7 MN/m^2 sometimes are needed to initiate grouting in a tight hole. It is important that the pressure build-up during injection is not too rapid. This can be controlled by adjusting the rate of pumping. A slower pumping rate gives a higher grout-take.

Pore-water pressures in clay soils should be monitored during grouting. In some instances slowing the pumping rate prevents the development of excess pore-water pressures. In others, grouting has to be interrupted to allow dissipation of the excess pore-water pressure. However, in certain cases drainage, in the form of band drains or sandwicks, must be provided.

The shape of the grout mass depends upon the character of the soil and the amount of grout-hole open at the time of injection. In uniform soils the shape is usually spherical or cylindrical, whereas it tends to be irregular in non-uniform soils. The size of the grout mass is influenced by the density, moisture content and mechanical properties of the soil, as well as the rate and pressure of injection. Masses of grout with diameters of 1 m or more are not uncommon (Warner, 1982). Grouting to depths in excess of 30 m has been used in compaction grouting.

9.5.4 Jet grouting

Jet grouting offers a means of forming an impermeable barrier (Coomber, 1986), as well as providing a means for supporting or underpinning structures. It can be used in all types of soils and poor soils in relatively inaccessible layers at depth can be replaced. The quantities of grout involved can be predicted with reasonable accuracy and close control can be exercised over the zones requiring treatment. In its simplest form jet grouting involves inserting an injection pipe into the soil to the required depth (Figure 9.13(a)). The soil is then subjected to a horizontally rotating jet of water. At the same time the soil is mixed with grout (cement or cement–bentonite) to form plastic soil–cement. The injection pipe is gradually raised. Replacement jet grouting involves

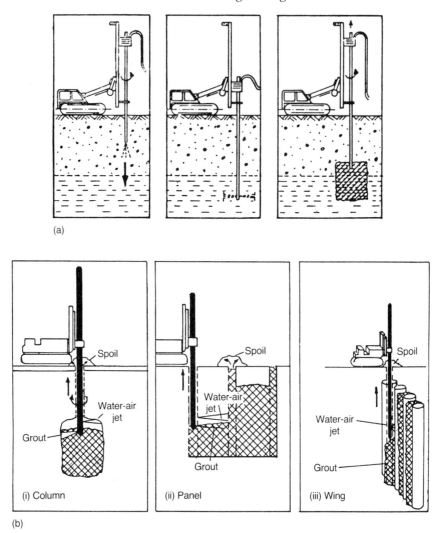

Figure 9.13 Jet grouting: (a) mixing grout with soil; (b) replacement jet grouting using compressed air and water to remove soil for simultaneous replacement by grout using column, panel and wing methods

the removal of soil from the zone to be treated by a high-energy erosive jet of water and air (Shibazaki and Ohta, 1982). The water and air are jetted under very high pressure from closely spaced nozzles at the base of a triple fluid phase drill pipe, the operation proceeding upwards from the base of a borehole. The grout is emplaced simultaneously. The soil that is removed is brought to the surface by air lift pressure. In granular soils some of the coarsest particles are not removed and are incorporated in the grout. This has proved to be one of the most promising means of deep stabilization of loess soils. Treatment has been completed successfully to depths in excess of 40 m.

There are three methods of replacement jet grouting: namely, column, panel or wing grouting (Figure 9.13(b)). However, the rows of columns or panels so formed cannot be regarded as free-standing or cantilever walls since they do not possess significant tensile strength and bending resistance. In column grouting, the columns are formed by rotating the jet pipe as it is raised (Coomber, 1986). A barrier is produced by overlapping the columns. In panel grouting, jetting takes place from the pipe in a single vertical plane as it is raised. Interconnecting panels form cut-offs of low permeability. Wing jetting uses two nozzles, without rotation, to form a fan-shaped grouted mass. The use of pre-bored guideholes, usually 150 mm in diameter, facilitates discharge of soil, helps to maintain verticality and provides a visual check on the continuity of adjacent grouted areas. Grouting can be terminated at a given level, without disrupting the overlying soil. The pipe is simply withdrawn, restricted grout flow only being maintained to top up the guidehole.

The grouts used must possess a low viscosity so that pumping can be maintained at the required rate. In addition, abrasive fillers such as sand must be avoided. Apart from these two factors the constituents of a grout mix can be varied to meet the requirement. For example, if low permeability is the principal criterion, then cement–bentonites (5% bentonite) normally are used.

Stocker and Zwicker (1987) described the formation of high-pressure injection columns to reduce differential settlement in the foundation for a highrise building. The clay soil varied in consistency from soft to stiff. Holes were drilled for the high-pressure injection columns by continuous flight augers to a depth exceeding 50 m. The holes were installed on a grid pattern at 2.3 m centres. The jet grouting pipes were lowered into the self-supporting uncased holes. A cement grout with a water cement ratio of 1 was used, the grouting pressure being 400 bars (40 MN/m^2), the pull-out speed was 0.2 m/min and the speed of rotation was 20 rev/min.

A major disadvantage in using jet grouting for pretreatment of the crowns of tunnels prior to excavation is that stress relief and loosening of the ground can occur since a void is created before the grout is injected. In order to overcome this, direct displacement injection has been introduced by removing the water erosion jet. The grout can be injected on its own or with a stream of air (Anon., 1991). Columns so formed contain more soil than water-jetted grout columns and therefore possess a lower strength, but there is less destressing around the columns. Also the friction between them and the soil is higher and the surrounding soil is more consolidated. Dugnani *et al.*

(1989) described such sub-horizontal jet grouting (Figure 9.14) to form a protective zone ahead of a tunnel face, thereby increasing the stability of the ground and reducing ground movements to a minimum.

A further innovation is reverse grouting in which grouting is carried out as the pipe enters the ground instead of as it is withdrawn. The technique requires higher capacity pumps than conventional jet grouting. In reverse grouting the nozzles are inclined downwards so that the jet has both horizontal and vertical components. Provided the soil conditions are suitable (a good rule of thumb is where the blowcount from a standard penetration test is less than 10) it is

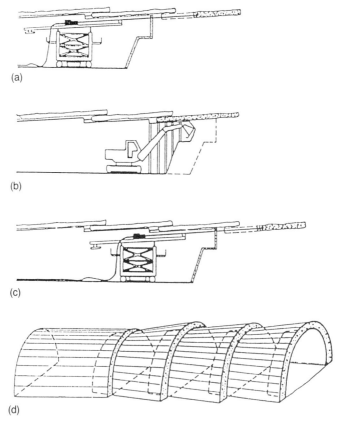

(a)

(b)

(c)

(d)

Figure 9.14 Sub-horizontal jet-grouting execution sequence for tunnels: (a) jet grouting ahead of the face; (b) excavation; (c) jet grouting at next stage; (d) geometrical scheme of treatment. Jet-grouted columns are formed ahead of the tunnel face, the overlapping sub-horizontal columns forming a protective shell around the excavation. The length of treatment is normally between 10 and 15 m but excavation stops 2 or 3 m before the end of the treated zone so as to leave sufficient support for roof stability (after Dugnani *et al.*, 1989)

possible to form a grout column directly using the vertical component of the high-energy jet to displace the soil at the nose of the pipe. The diameter of a column can be increased from 0.6 (conventional jet grouting) to 1.5 m when grout alone is injected and to 7 m when grout is injected with a stream of air.

9.5.5 Some other aspects of grouting

Construction of horizontal grouted diaphragms, which are connected to impermeable vertical cut-offs, are required when excavations extend below the water table and groundwater lowering techniques, for one reason or another, cannot be used and an impermeable formation does not exist at a suitable depth for the vertical cut-off to be keyed into (Tausch, 1985). Such diaphragms should be 1–2 m in thickness. Because of uplift pressure, they must be constructed at a level significantly below the proposed base of the excavation in order to prevent blow-out into the excavation. The grout may be emplaced at the required depth by a *tube-à-manchette*. Grout-holes are spaced at 1.1–1.5 m, depending upon ground conditions, and are usually set out in a triangular pattern (Figure 9.15). The grout injected from each hole overlaps with that injected from the adjacent holes. Thick diaphragms are formed by grouting in more than one layer.

The reduction of percolation of water through the ground can be accomplished by forming a grout curtain which involves drilling lines of

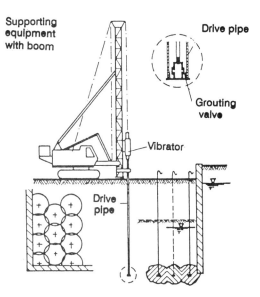

Figure 9.15 Construction of horizontal grout membrane

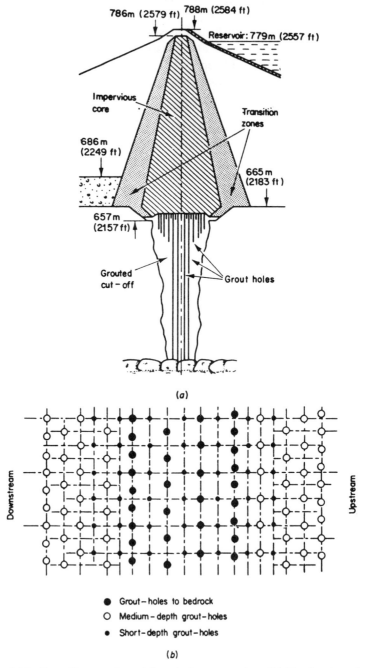

Figure 9.16 Serre Poncon dam: (a) typical cross-section; (b) distribution of grout holes in the central part of the grouted cut-off

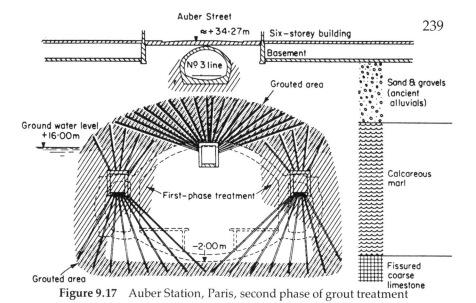

Figure 9.17 Auber Station, Paris, second phase of grout treatment

grout-holes in the ground to create a barrier or cut-off against excessive seepage (Figure 9.16). The spacing of grout-holes varies from about 1 to 2.5 m apart and grouting takes place either from an open-pipe or a *tube-à-manchette*. Once the standard of permeability has been decided for the whole or a section of a grout curtain, it is achieved by split spacing or closure methods in which primary, secondary, tertiary, etc., sequences of grouting are carried out until water tests in the grout-holes approach the required standard. In multiple row curtains the outer rows should be completed first, thereby allowing the innermost row to effect closure on the outer rows. The upstream row should be the tightest row, tightness decreasing downstream. Single-row curtains are usually constructed by drilling alternative holes first and then completing the treatment by intermediate holes. Ideally a grout curtain is taken to a depth where the requisite degree of tightness is available naturally.

Grouting in water-bearing ground is frequently carried out ahead of a tunnel face in order to reduce the quantity of water entering the tunnel to readily manageable amounts (Tan and Clough, 1980). In particular, difficulties frequently arise when tunnels are excavated beneath rivers which contain buried channels, especially those occupied by sands and gravels. Such grouting has been referred to as *aureole* or *umbrella* grouting and involves drilling grout-holes in advance of the tunnel heading which fan out to form a series of concentric grouted cones (Figure 9.17). The tunnel is then excavated through the grouted zone. The length of individual holes depends upon the type of ground on the one hand and the quantity and pressure of water on the other. In most cases, 9 m should prove sufficient. The type of grout used to form the aureole is governed by the ground conditions.

10

Soil stabilization

The objectives of mixing additives with soil are to improve volume stability, strength and stress–strain properties, permeability, and durability. The development of high strength and stiffness is achieved by reduction of void space, by bonding particles and aggregates together, by maintenance of flocculent structures, and by prevention of swelling. The permeability is altered by modification of pore size and distribution. Good mixing of stabilizers with soil is the most important factor affecting the quality of results. The two most commonly used stabilizers are cement and lime.

10.1 CEMENT STABILIZATION

The addition of small amounts of cement, that is, up to 2%, modify the properties of a soil, while large quantities cause radical changes in these properties. The amount of cement needed to stabilize soil has been related to the durability requirement; put another way, a minimum unconfined compressive strength of 2.8 MN/m^2, after curing at a constant temperature (25°C) and moisture content for seven days, has been widely used. In fact cement contents may range from 3 to 16% by dry weight of soil, depending on the type of soil and properties required (Table 10.1). Generally as the clay content of a soil increases, so does the quantity of cement required.

Any type of cement may be used for soil stabilization but ordinary Portland cement is most widely used. The two principal factors that determine the suitability of a soil for stabilization with ordinary Portland cement are, firstly, whether the soil and cement can be mixed satisfactorily and, secondly, whether, after mixing and compacting, the soil–cement will harden adequately. Rapid-hardening cement with extra calcium is used in organic soils, and a retarded cement will tolerate construction delays. Sulphate-resisting cements are rarely suitable.

Table 10.1 Typical cement requirements for various soil types (after Anon., 1990(d))

Unified soil classification	Typical range of cement requirement,* (% by wt)	Typical cement content for moisture-density test (ASTM D 558),† (% by wt)	Typical cement contents for durability tests (ASTM D 559 and D 506),‡ (% by wt)
GW, GP, GM, SW, SP, SM	3–5	5	3–5–7
GM, GP, SM, SP	5–8	6	4–6–8
GM, GC, SM, SC	5–9	7	5–7–9
SP	7–11	9	7–9–11
CL, ML	7–12	10	8–10–12
ML, MH, CH	8–13	10	8–10–12
CL, CH	9–15	12	10–12–14
MH, CH	10–16	13	11–13–15

* Does not include organic or poorly reacting soils. Also, additional cement may be required for severe exposure conditions such as slope protection.

† ASTM D 558 (1992) *Standard Test Method for Moisture-Density Relations of Soil–Cement Mixtures*, American Society for Testing Materials, Philadelphia.

‡ ASTM D 559 (1982) *Standard Methods for Wetting and Drying Tests of Compacted Soil-Cement Mixtures*, American Society for Testing Materials, Philadelphia. ASTM D 506 (1982) *Standard Methods for Freezing and Thawing Tests of Compacted Soil–Cement Mixtures*, American Society for Testing Materials, Philadelphia.

10.1.1 Types of soil

Any type of soil, with the exception of highly organic soils or some highly plastic clays, may be stabilized with cement. Although particles larger than 20 mm diameter have been incorporated in soil–cement, a maximum size of 20 mm is preferable since this allows a good surface finish. At the other extreme, not more than about 50% of the soil should be finer than 0.18 mm. Granular soils are preferred since they pulverize and mix more easily than fine-grained soils and so result in more economical soil–cement as they require less cement. Typically soils containing between 5 and 35% fines yield the most economical soil–cement. As the grain size of granular soils is larger than that of cement, the individual grains are coated with cement paste and bonded at their points of contact.

The particles in cohesive soils are much smaller than cement grains and, consequently, it is impossible to coat them with cement. In practice, cohesive soils are broken into small fragments which are coated with cement and then compacted. The hydration products formed after short periods of ageing are largely gelatinous and amorphous which, with time, harden due to gradual desiccation. With further curing, poorly ordered varieties of hydrated calcium silicate and hydrated calcium aluminate develop. Ultimately the hydrated cement forms a skeletal structure, the strength of which depends on the size of the fragments and amount of cement used. A secondary change occurs in clay soils due to the free lime in the cement, which reacts with the clay particles, making the soil less cohesive. However, clay balls may form when the plasticity index exceeds 8%. Where the soil–cement is exposed to weathering, the clay balls break down, which weakens the soil–cement. In such instances the US Bureau of Reclamation requires that clay balls larger than 25 mm diameter are removed (Anon., 1986b).

It is difficult to mix dry cement into heavy clays and high amounts of cement have to be added to bring about appreciable changes in their properties. Indeed, clay soils with liquid limits exceeding 45% and plasticity indices above 18% are not usually subjected to cement stabilization (Croft, 1968). Heavy clays can, however, be pretreated with 2–3% cement or, more frequently, with hydrated lime. This reduces the plasticity, thereby rendering the clay more workable. After curing for 1–3 days the pretreated clay is stabilized with cement.

Furthermore, the suitability of a clay soil for cement stabilization is controlled by its texture, and chemical and mineralogical composition. Both kaolinite and well-crystallized illite have little or no effect on the hydration and hardening process of cement stabilization. By contrast, the expansive clay minerals, depending upon their relative activities, may have a profound influence on the hardening of cement (Bell, 1976). For instance, the affinity of montmorillonite for lime reduces the pH value of the aqueous phase, and because of the deficiency in lime the cementitious products developed during curing are inferior to those of non-expansive clays. This means that the strengths developed are lower, and unless enough cement is added to supply the free lime requirement to promote hardening, the properties of the clay are not enhanced. Up to 15% of cement has to be added to montmorillonitic clays to modify them significantly.

The behaviour of weathered minerals such as degraded illites, chlorites and vermiculites can be similar to that of montmorillonite. Gibbsite, with its high response to lime, may retard stabilization of certain lateritic soils. In general, lime will be more suitable than cement for soils containing these components.

Organic matter and excess salt content, especially sulphates, can retard or prevent hydration of cement in soil–cement mixtures. In fact, soils containing more than 2% organic material are usually considered unacceptable (Anon., 1990(d)) and soils with pH values of less than 5 are unsuitable for economic stabilization.

Organic matter retards the hydration of cement because it preferentially absorbs calcium ions. However, the addition of calcium chloride or hydrated lime can provide a source of calcium and, consequently, may enable some of these soils to be treated. In addition, Kuno *et al.* (1989) found that a mixture of ordinary Portland cement and 10% (by weight) of gypsum when added to organic soils gave better results in terms of strength development than did cement alone.

The disintegration of cement (or lime) stabilized soils, due to sulphate attack, only occurs when the soil has an appreciable clay fraction and when there is an increase in moisture content above that at the time of compaction. Sherwood (1957) maintained that there is a risk of deterioration of clay–cement mixtures when the content of SO_3 and SO_4 in the soil is 0.2% and 0.5% or more, respectively, or if the SO_3 content in groundwater exceeds 300 mg/l. He found that sulphate-resistant cement was no better than ordinary Portland cement for stabilizing clay soils containing sulphates.

10.1.2 Mixing and compaction

In order to achieve a uniform material with minimum cement content, good standards of mixing are required. Two basic methods of mixing are employed. In the premix method all the soil is obtained from a borrow pit. Then it is batched, by weight or volume, into a mixer. After mixing, material is transported to the site, where it may be spread by hand, grader, stone spreader, concrete spreader or bituminous paver (Figure 10.1) Compaction is usually by roller. Vibratory rollers are suitable for granular materials, and dead weight rollers for cohesive materials. Rubber-tyred rollers appear to operate efficiently on a wide range of materials.

In the mix-*in-situ* method, mobile mixers are employed to mix the materials in place on site. If the soil on site is used, savings accrue from a reduction in the volume of earthworks required. However, in some cases it is more economic to treat imported materials which are spread, compacted and levelled, prior to cement spreading and mixing. The machines used for mixing vary from agricultural rotary tillers to large purpose-built units (Figure 10.2). The smaller machines are limited to processing depths of 150–200 mm, while some of the larger machines can process depths in excess of 400 mm. Standards of mixing are improved, especially in cohesive materials, if rotary tillers are used to pulverize the soil into small fragments prior to adding cement. It is also often necessary to add water at this stage. On small jobs bagged cement may be spread manually, but this is not economic on medium to large size jobs where mechanical spreaders are used. After mixing in the cement and any additional water needed to reach the optimum moisture content, compaction and grading to final level is carried out. Finally, the processed layer is covered with a waterproof membrane, commonly

Figure 10.1   Premixed soil–cement being laid as a sub-base for a road. Note the cover of bitumen in the foreground which acts as a waterproof membrane

bitumen emulsion, to prevent drying out and to ensure cement hydration.

10.1.3 Properties of cement-stabilized soils

The properties developed by compacted cement-stabilized soils are governed by the amount of cement added on the one hand and compaction on the other. With increasing cement content the strength and bearing capacity increase, as does the durability to wet–dry cycles. The permeability generally decreases but tends to increase in clayey soils. Granular soils may become more prone to shrinkage, whereas the swellability of clay soils is reduced (Table 10.2).

  The density achieved is largely a function of compactive effort, soil texture and, in the case of clay soils, the type of clay minerals present, which determine the soil moisture response. Adequate compaction is essential for successful stabilization but prolonged delays between mixing and compaction reduce the maximum density attainable (Figure 10.3). The addition of cement produces small increases in the compacted densities of both kaolinitic and illitic clay soils, but not those containing

Figure 10.2 Autograder producing mix-in-place stabilized soil (courtesy Bomag Ltd)

montmorillonite; in fact the latter gives rise to small reductions in compaction densities.

Soil–cement undergoes shrinkage during drying and soil–cement made with different types of soils shows different crack patterns. For example, soil–cement made with clay develops a higher total shrinkage but crack widths are smaller and more closely spaced than in soil–cement made with sand. The development of shrinkage cracks can be reduced by keeping the surface of the soil–cement moist beyond the normal period of curing and placing it at slightly below optimum moisture content.

The strength of soil–cement tends to increase in a linear manner with increasing cement content, but in different soils it increases at different rates (Figure 10.4(a)). Increased pulverization increases the strength of soil–cement. Ideally the degree of pulverization, if the larger stone particles are excluded, should break down to parent soil so that 80% of the particles are less than 5 mm and mixing should be uniform. A

Table 10.2 Typical average properties of soil–cement and soil–lime mixtures (after Ingles and Metcalf, 1972)

(a) Typical mean* properties of soil–cement†

Soil type (unified classification)	Compressive strength (MN/m²)	Young's Modulus, E (MN/m²)	CBR	Permeability (m/s)	Shrinkage	Comments
GW, GP, GM, GC, SW	6.5	2×10^4	>600	Decreases ($\approx 2 \times 10^{-7}$)	Negligible	Too strong; liable to wide spaced cracks‡
SM, SC	2.5	1×10^4	600	Decreases	Small	Good material
SP, ML, CL	1.2	5×10^3	200	Decreases ($\approx 1 \times 10^{-9}$)	Low	Fair material
ML, CL, MH, VH	0.6	2.5×10^3	<100	Increases	Moderate	Poor material
CH, OL, OH, Pt	<0.6	1×10^3	<50	Increases ($\approx 1 \times 20^{-11}$)	High	Difficult to mix; needs excessive cement

* Variations of approximately 50% around the mean may be expected.
† Values shown are at 10% cement content.
‡ Good material if *less cement* is used.

(b) Typical mean* properties of soil–lime†

Soil type (unified classification)	Compressive strength (MN/m²)	Young's Modulus, E (MN/m²)	CBR	Permeability (m/s)	Linear shrinkage (%)	Plasticity index	Comments
GW, GP, GM, GC, SW, SP	≤0.3	–	75	Increases ($\geq 10^{-7}$)	Nil	Non-plastic	Suitable only for plasticity reduction
SM, SC	1.1	$<1 \times 10^2$	50	Increases	Very low	<5	Poor to fair material
ML, CL, MH, VH	2.5	2×10^4	30	Increases	5	10	Good material
CH	3.5	1×10^3	25	Increases ($\leq 10^{-10}$)	10	20	Good effect, fair to good material
OL, OH, Pt	≤1.0	1×10^2	≤10	–	–	15	Not suitable *per se*‡

* Variations of approximately 50% around the mean may be expected.
† Values shown are at the additive level optimum for the respective soil types.
‡ Results may be improved by admixture of the lime with gypsum.

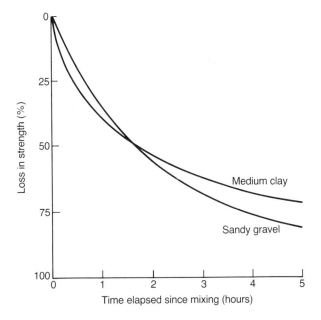

Figure 10.3 Loss in strength due to delay in compaction for two soils stabilized with 10% cement; standard compaction (from Ingles and Metcalf, 1972)

lengthy period of mixing brings about partial hydration of the cement with a resultant loss of strength at constant density. If compaction is delayed the cement begins to hydrate and therefore the soil–cement begins to harden. As a result the mixture becomes more difficult to compact. Compaction should be completed within two hours of mixing. The strength of soil–cement gradually increases as the time taken in curing increases (Figure 10.4(b)). Also, the higher the temperature, the more rapid is the gain of strength. Soil–cement will harden in cold weather providing the temperature does not fall below 0°C. Excessive drying increases strength but tends to crack the soil–cement. By contrast, strength is reduced by soaking (Figure 10.5). This is particularly the case with clayey soils. Typical ranges of 7- and 28-day unconfined compressive strengths for soaked soil–cement specimens are given in Table 10.3. It is recommended by Anon. (1990(d)) that specimens are soaked before being tested since soil–cement may become intermittently or permanently saturated during its service life and it possesses lower strength under saturated conditions.

Values of Young's modulus vary between 140 and 20000 MN/m^2 depending on soil type and increase with increasing cement content. Plastic deformation occurs on cyclic loading, and under such conditions failure may occur at 60–70% of the ultimate failure strength. When

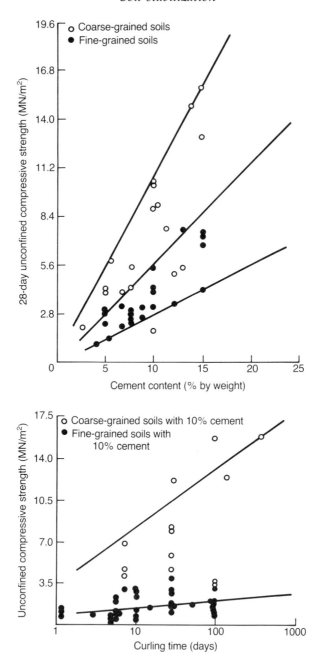

Figure 10.4 (a) Relationship between cement content and unconfined compressive strength for soil–cement mixture; (b) effect of curing time on unconfined compressive strength of some soil–cement mixtures (from Anon., 1990(d))

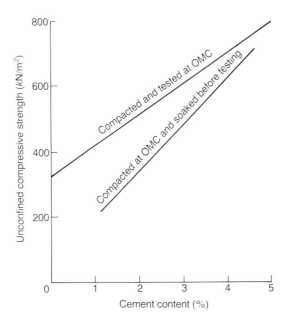

Figure 10.5 Effect of soaking on strength (after Ingles and Metcalf, 1972)

subjected to constant loading, soil–cement undergoes creep.

The permeability of most soils is reduced by the addition of cement. In multiple-lift construction, the permeability along horizontal surfaces is generally greater than along vertical; for instance, Nussbaum and Colley (1971) found that permeabilities for flow parallel to the plane of compaction of an embankment were 2–20 times greater than in the plane normal to the direction of compaction.

Table 10.3 Ranges of unconfined compressive strengths of soil–cement (after Anon., 1990(d))

	Soaked compressive strength* (MN/m^2)	
Soil type	*7-day*	*28-day*
Sandy and gravelly soils: Unified groups GW, GC, GP, GM, SW, SC, SP, SM	2·07–4·14	2·76–6·90
Silty soils: Unified groups ML and CL	1·72–3·45	2·07–6·21
Clayey soils: Unified groups MH and CH	1·38–2·76	1·72–4·14

* Specimens moist-cured 7 or 28 days, then soaked in water prior to strength testing.

10.1.4 Applications of soil–cement

The principal use of soil–cement is as a base material underlying pavements. One of the reasons soil–cement is used as a base is to prevent pumping of fine-grained subgrade soils into the pavement above. The thickness of the soil–cement base depends upon subgrade strength, pavement design period, traffic and loading conditions and thickness of the wearing surface. Frequently, however, soil–cement bases are around 150–200 mm in thickness.

Soil–cement has been used to afford slope protection to embankment dams, soil–cement made from sandy soils giving a durable erosion-resistant facing. Soil–cement has also provided slope protection for canals, river banks, spillways, highway and railway embankments and coastal cliffs. Where slopes are exposed to moderate to severe wave action or rapidly flowing water, the soil–cement generally is placed in horizontal layers 150–225 mm thick and 2–3 m wide adjacent to the slope, that is, as 'stairstep slope protection'. In situations where conditions are less severe, a layer of soil–cement 150–225 mm thick may be placed parallel to the slope of the face.

In addition to water storage reservoirs, soil–cement has been used to line waste-water treatment lagoons, sludge-drying beds, ash-settling ponds and sanitary landfills. The soil–cement linings are commonly 100–150 mm thick. Tests sponsored by the US Environmental Protection Agency (Anon., 1983) showed that when soil–cement was exposed to leachate from sanitary landfills, the soil–cement hardened considerably and became less permeable. Tests carried out with hazardous wastes indicated that no seepage had occurred through the soil–cement after 2.5 years of exposure. Nonetheless, waste materials should be tested to determine their compatibility with soil–cement. These tests did not include exposure of soil–cement to acid wastes.

Soil–cement has been used as massive fill replacement to provide uniform support to foundations where inadequate bearing capacity was available. Such replacement fills have ranged up to 5.5 m in thickness (Dupas and Pecker, 1979).

Soil–cement cushions have been widely used in Bulgaria for the support of structures on collapsible loess soils (Minkov *et al.*, 1980). The soil–cement cushion is constructed using loess from the building site. About 3–7% cement is mixed with the loess and is then compacted in layers between 150 and 200 mm thick. The cushion is usually 1–1.5 m in thickness, but on some occasions has exceeded 3 m thick. Loess–cement cushions also have been used in conjunction with heavy tamping.

10.1.5 Plastic soil–cement

On occasions, soil and cement have been mixed at water contents higher than the consistency of wet concrete and then placed without being compacted. Such material has been termed plastic soil–cement and a strong durable product can be obtained using most types of soil except very clayey or organic types. Cohesionless soils usually form a stronger plastic soil–cement than cohesive soils. Plastic soil–cement made from silty sand mix may require about 30% water (this is about twice the desirable water content that would have been used if the material was compacted with a roller).

In order to achieve comparable strengths, plastic soil–cement needs the addition of more cement than normal cement-stabilized soil, and its density is lower. Nevertheless, a compressive strength of 3.5–7.0 MN/m^2 can be attained after 28 days with most soils. The strength continues to increase as the material ages, so that after a year it frequently has doubled. Plastic soil–cement, when cured in moist or saturated ground, tends not to shrink even when it contains a high percentage of clay.

Plastic soil–cement can be pumped into place to form backfill or tremied into slurry trenches to form impervious walls. It also has been used to provide thin linings for irrigation canals.

Plastic soil–cement can be mixed *in situ* to form shallow piles (9–12 m in depth) in soils that are not excessively difficult to dig. Cement is injected into the soil by a special drilling and mixing bit as the hole is advanced, as well as when the bit is raised and rotated. The resulting columns of plastic soil–cement have not only proved successful as structural piles but overlapping piles have been constructed to form underground walls. Normally between 35 and 50% of liquid cement by volume is added to the volume of the completed soil–cement pile. As a significant quantity of water flows from the soil–cement into neighbouring soil, the water/cement ratio of the pile is reduced. This is largely responsible for the high strengths attained by some piles. Also, the curing conditions are ideal since the soil–cement is surrounded by either moist or saturated soil at a desirable temperature.

Another method of forming a mixed *in-situ* soil–cement pile involves drilling the hole using water flush, then, as the drill stem is lifted, injecting cement grout horizontally into the hole from jets in the bit (see section 9.5.4 in Chapter 9). The mixing is achieved by the turbulent action created by jetting. The pile so formed has a diameter some three to four times that of the original hole.

10.2 LIME STABILIZATION

Lime stabilization refers to the stabilization of soil by the addition of burned limestone products, either calcium oxide (i.e. quicklime, CaO) or calcium hydroxide, $Ca(OH)_2$. On the whole, quicklime appears a more effective stabilizer of soil than hydrated lime. Moreover, when quicklime is added in slurry form it produces a higher strength than when it is added in powder form. The process is similar to cement stabilization except that lime stabilization is applicable to much heavier clayey soils and is less suitable for granular materials. In fact, the addition of lime has little effect on soils that contain either a small clay content or none at all. It also has little effect on highly organic soils.

Lime usually reacts with most soils with a plasticity index ranging from 10 to 50%. Those soils with a plasticity index of less than 10% require a pozzolan for the necessary reaction with lime to take place, fly ash being commonly used. Other pozzolans used for the enhancement of lime stabilization include blast furnace slag and expanded shale. Lime is particularly suited to the stabilization of heavy clays and may be more effective than cement in clayey gravels. Lime stabilization of heavy clays gives the soil a more friable structure, which is easier to work and compact, although a lower maximum density is obtained. The reaction of lime with montmorillonitic clays is quicker than with kaolinitic clays; in fact, the difference may amount to a few weeks. A silica surface, however, should not be considered 'available' if it is bound to a similar surface by ions which are not readily exchangeable. Accordingly, illite and chlorite, although attacked, are much less reactive than montmorillonite.

When lime is used to stabilize clay soil it forms a calcium silicate gel which coats and binds lumps of clay together and occupies the pores in the soil. Reaction proceeds only while water is present and able to carry calcium and hydroxyl ions to the surfaces of the clay minerals (that is, while the pH value is high). Consequently, reaction ceases on drying and very dry soils do not react with lime.

The quantity of lime added should ideally be related to the clay mineral content as the latter is needed for reaction. Ingles and Metcalf (1972) suggested that the addition of up to 3% of lime would modify silty clays, heavy clays and very heavy clays, while 3–4% was required for the stabilization of silty clay, and 3–8% was proposed for stabilization of heavy and very heavy clays. They further suggested that a useful guide is to allow 1% of lime (by weight of dry soil) for each 10% of clay in the soil. More exact prescriptions can usually be made after tests at and slightly each side of the guide value.

Like cement stabilization, lime stabilization tends not to be very effective in organic soils since the organic matter retards hydration. However, Kuno *et al.* (1989) showed that when 20% (by weight) of gypsum was added to quicklime or hydrated lime, then this mixture could be used to stabilize organic soils as long as they did not possess excessively high natural moisture contents (Figure 10.6).

The amount of water used in lime stabilization is dictated by the requirements of compaction. However, if quicklime is used, then extra

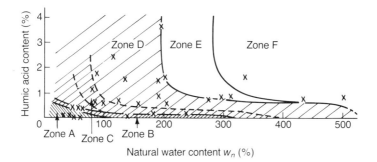

| | Appropriateness of *cementitious* admixtures | | | | |
Soil classifi-cation	Slaked lime	Quicklime	Slaked lime + gypsum	NPC	NPC + gypsum
Zone A	●	●	●	●	●
Zone B	○	○	●	●	●
Zone C	x	○	○	●	●
Zone D	x	x	x	●	●
Zone E	x	x	x	x	○
Zone F	x	x	x	x	x

● . . . . . Short-term strength increase excellent.
○ . . . . . Short-term strength increase not so good, but develops long-term strength.
x . . . . . . Strength does not develop in long term.

Figure 10.6 Soil classification zones based on the natural water content and humic acid content with guidelines for selecting admixtures on the basis of the soil classification (after Kuno *et al.*, 1989)

water may be necessary in soils with less than 50% moisture content to allow for the very rapid hydration process. Furthermore, because the lime–soil reaction involves exsolution, it is inhibited if the water content of the soil is too low. Hence moist curing is always desirable.

Mixing is important, and if mixing is delayed after the lime has been exposed to air, then carbonation of the lime will reduce its effectiveness. Therefore, it is desirable that mixing be effected as soon as possible and certainly within 24 hours of exposure to air.

10.2.1 Properties of lime-stabilized soils

In most cases the effect of lime on the plasticity of clay soils is more or less instantaneous. In other words, the plasticity is reduced (this is brought about by an increase in the plastic limit and reduction in the

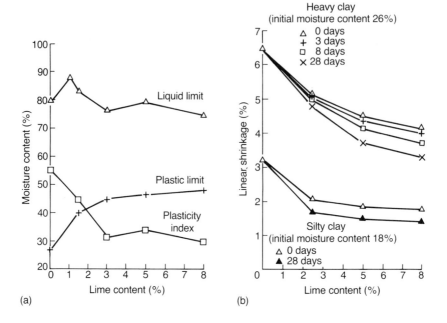

Figure 10.7 (a) Influence of the addition of lime on the plastic limit, liquid limit and plasticity of clay of high plasticity. (b) Influence of lime on the linear shrinkage of heavy and silty clay (from Bell, 1988)

liquid limit of the soil), as is the potential for volume change (Figure 10.7). For example, tests carried out by the US Bureau of Reclamation indicated that the addition of 4% lime reduced the plasticity index of a clay from 47 to 12% and increased the shrinkage limit from 7 to 26% (Anon., 1975). In kaolinitic clay soils, however, lime treatment at times increases the plasticity index.

The addition of lime to clayey soils increases the optimum moisture content and reduces the maximum dry density for the same compactive effort (Figure 10.8). The significance of these changes depends upon the amount of lime added and the amount of clay minerals present. As lime treatment flattens the compaction curve, a given percentage of the prescribed density can be achieved over a much wider range of moisture contents so that relaxed moisture control specifications are possible.

The strength of soil–lime mixtures depends on several factors such as soil type and the type and amount of lime added. For example, montmorillonitic clays give lower strengths with dolomitic limes than with high-calcium or semi-hydraulic limes (Bell, 1988). Kaolinitic clays, on the other hand, yield the highest strengths when mixed with semi-hydraulic limes and the lowest strengths are obtained with high-calcium limes.

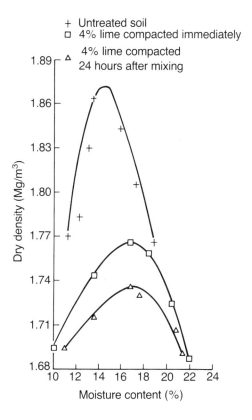

Figure 10.8 Influence of the addition of lime on the compaction curves of clay soil (from Bell, 1988)

Soil mixed with low lime content attains a maximum strength in less time than that to which a higher content of lime has been added. Strength does not increase linearly with lime content and in fact excessive addition of lime reduces strength (Figure 10.9(a)). This decrease is because lime itself has neither appreciable friction nor cohesion. The optimum lime content tends to range from 4.5 to 8%, the higher values being required for soils with higher clay fractions. Curing time is another factor influencing strength of lime-stabilized soil (Bell and Couthard, 1990), a steady gain in strength over months is characteristic (Figure 10.9(b)). Higher temperatures accelerate curing, and this gives rise to higher strengths (Figure 10.9(c)). The soil–lime reaction is retarded or may cease once temperatures fall below 4°C. Hence, strength development more or less ceases with the onset of cold weather, and strength loss because of cyclic freezing and thawing is cumulative throughout the winter. Residual strength at the end of the

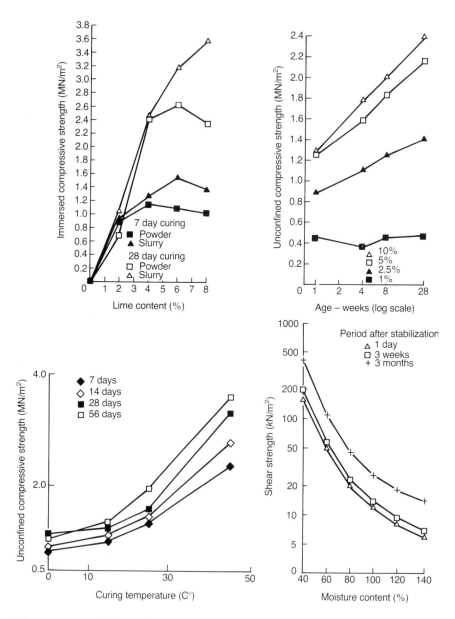

Figure 10.9   (a) Effect of the addition of lime, in slurry and powder form, on the strength of silty clay; (b) Influence of curing time on the unconfined compressive strength of clay soil of high plasticity treated with different amounts of lime; (c) Influence of curing temperature and time on strength of clay of very high plasticity stabilized with 5% addition of lime; (d) Influence of addition of 4% lime on the shear strength of clay soil with different moisture contents (from Bell, 1988)

freeze–thaw period consequently must be sufficient to guarantee the integrity and stability of the soil–lime layer.

Yet another factor influencing the strength of lime-stabilized soils is their natural moisture content, the strength decreasing with increasing moisture content. Even three months after stabilization, the shear strengths of soils with high saturated moisture contents remain low (Figure 10.9(d)). On the other hand, soil–lime mixtures compacted at moisture contents above optimum, after brief periods of curing, attain higher strengths than those compacted with moisture contents less than optimum. This is probably because the lime is more uniformly diffused and occurs in a more homogeneous curing environment. The strength of soils treated at or below optimum moisture content usually can be improved by spraying each lift with water after compaction.

Al-Rawi and Awad (1981) indicated that the permeability of a clayey soil is increased when treated with lime as the latter causes flocculation of the soil. The higher the clay fraction, the more the permeability of the soil–lime mixture increases. They found that those soils compacted on the dry side of optimum moisture content developed a higher permeability than those compacted wet of optimum. They also identified particular moulding moisture contents which gave minimum permeability for soil–lime mixtures irrespective of the amount of lime added. However, with increasing age the permeability declines, this being explained by Brandl (1981) as due to the increased production of gelatinous reaction products which occupy an increasing amount of void space and eventually harden.

10.2.2 Uses of lime stabilization

The principal use of the addition of lime to soil is for subgrade and sub-base stabilization and as a construction expedient on wet sites where lime is used to dry out the soil. As far as lime stabilization for roadways is concerned, stabilization is brought about by the addition of between 3 and 6% lime (by dry weight of soil).

Subgrade stabilization involves stabilizing the soil in place or stabilizing borrow materials which are used for sub-bases. After the soil, which is to be stabilized, has been brought to grade, the roadway should be scarified to full depth and width and then partly pulverized. A rooter, grader-scarifier and/or disc harrow for initial scarification, followed by a rotary mixer for pulverization, is employed.

Dry hydrated lime can be spread uniformly by a mechanical spreader or from bags emptied in piles and levelled off by a drag pulled by a tractor. Some idea of the amount of lime that is required can be obtained from Figure 10.10. After sprinkling dry lime with water, preliminary mixing is required to distribute the lime thoroughly throughout the soil

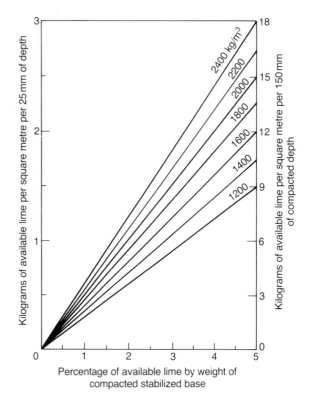

Figure 10.10   Rate of application of lime (after Anon, 1982)

to the proper depth and width and to pulverize the soil to −50 mm. During mixing, water is added to bring the soil slightly above the optimum moisture content.

   Lime slurries of varying concentrations, depending on the percentage of lime required and the optimum moisture content, are also applied to the soil. A typical mix consists of 1 tonne of lime to 2500 litres of water, which produces approximately 31% lime solution. At higher concentrations there is difficulty in pumping and handling the slurry spray bars. Forty per cent is a maximum pumpable slurry. Where low lime percentages are required, the blend may be cut down to 1 tonne of lime to 3000–3500 litres of water. Since lime in the form of slurry is much less concentrated than dry lime applications, usually two or more passes are required to provide the specified amount. Although the slurry method promotes more uniform distribution of lime, it proves disadvantageous on wet soils. Its use is generally restricted to projects which require smaller amounts of lime (4% or under) since, at higher percentages, so much water is required to carry the lime in solution that the soil would exceed optimum moisture content most of the time.

After initial mixing the lime-treated soil should be lightly compacted to reduce carbonation and evaporation loss. Initial curing takes 24–48 hours, water being added to maintain as near optimum mixture conditions as possible. Where necessary the soil–lime should be remixed so that it is friable enough for the required degree of pulverization to be achieved.

Soil–lime mixtures should be compacted to high density in order to develop maximum strength and stability. This necessitates compacting at or near the optimum moisture content. Granular bases are generally compacted as soon as possible after mixing. However, delays of up to two days are not detrimental, especially if the soil is now allowed to dry out. Clay subgrades can be compacted soon after final mixing although delays of up to four days are not injurious. When longer delays (two weeks or more) cannot be avoided, it may be necessary to incorporate a small amount of extra lime into the soil (for example, 0.5%) to compensate for losses due to carbonation.

Generally, a five- to seven-day curing period is required. Two types of curing are employed. Firstly, the soil is kept damp by sprinkling, with light rollers being used to keep the surface knitted together. Secondly, in membrane curing, the stabilized soil is either sealed with one application of asphalt emulsion within one day of final rolling or primed with increments of asphalt emulsion applied several times during the curing period.

Heavy costs can be incurred when construction equipment and transport become bogged down on site due to heavy rainfall turning clayey ground into mud. In these circumstances, an economical method of drying out the top layers of soil is essential. This can be achieved by the use of quicklime in granular form, it combining with soil moisture to produce hydrated lime. In the process, heat is generated which helps the soil dry out.

Lime stabilization is used in embankment construction for roads, railways, earth dams and levees to enhance the shear strength of the soil. In retaining structures it is used primarily to increase the resistance to water, either external or internal. For example, lime has been used to stabilize small earth dams constructed of dispersive soil and so avoid piping failure. Lime has also been used to stabilize low-angled slopes, a surface layer of soil about 150 mm thick being mixed in place.

Lime stabilization of clay soils, especially expansive clay soils, can minimize the amount of shrinkage and swelling they undergo. Hence, such treatment can be used to reduce the number and size of cracks developed by buildings founded on suspect clay soils. Lime stabilization may be applied immediately beneath strip footings for light structures. The treatment can be better applied as a layer below a raft in order to overcome differential movement.

Soil stabilization

Figure 10.11 Lime slurry injection method (coutesy of Woodbine Corporation, Fort Worth, Texas)

The lime slurry pressure injection method involves pumping hydrated lime slurry under pressure into expansive clay soils (Figure 10.11). The method has been used to improve the bearing capacity and reduce differential settlements, to stabilize failed embankment slopes, to minimize subgrade pumping beneath railways and as remedial treatment to stabilize soils beneath structures (Joshi *et al.*, 1981). Usually a depth of 2.1 m is sufficient to emplace lime slurry below the critical zone of changes in moisture content, although depths of up to 40 m can be stabilized. As the slurry is pumped under pressure into the soil, it follows the path of least resistance, moving along fissures and bedding planes and fractures created by pumping the slurry. Injection pressures usually range between 345 and 1380 kN/m², depending on soil conditions, injection taking place until refusal. If the points of injection are spaced at 1.5 m centres, the lime slurry forms a network of horizontal sheets interconnected by vertical veins. After injection, the network of soil–lime thickens as a result of calcium ion exchange. This means that there is a gradual improvement in the swell–shrink behaviour of the soil. Rogers (1991) has described the use of the lime slurry injection method to stabilize slopes.

Lime columns, 0.5 m in diameter, have been constructed in sensitive clay soils, to depths of 10 m, by means of a tool reminiscent of a giant

Figure 10.12 'Eggbeater' forming lime column (courtesy of Linden-Alimak)

'eggbeater' (Figure 10.12). The mast and the rotary table of the tool are usually mounted on a front wheel-loader. A container is attached to the loader to store the quicklime. The tool is screwed into the soil to the required depth, then the rotation is reversed and unslaked lime is forced into the soil, by compressed air, through openings placed just above the blades of the mixing tool. The amount of lime used approximates to 5–8% of the dry weight of the soil. The tool is extracted at a rate about one-tenth of its screwed-in rate so that lime is thoroughly mixed with the soil. Lime columns can be used to support light structures instead of piles (Holm *et al.*, 1981).

The bearing capacity of a lime column 0.5 m in diameter normally varies between 50 and 500 kN, depending on the type and amount of lime added. The strength of a lime column and the increase in strength after installation are affected by the curing conditions, the confining conditions and pore pressure may also affect column strength (Ahnberg *et al.*, 1989). The columns reduce both the total and differential settlements. Another application is around structures supported by piles, where the lime columns can reduce the negative skin friction and the lateral displacement of the piles due to creep of the surrounding soil. Interlocking lime columns have been used around excavations to form a less rigid block of stabilized soil (Kujala *et al.*, 1985; Saitoh *et al.*, 1985).

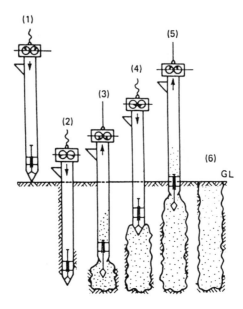

Figure 10.13   Installation of lime piles

Furthermore, lime columns bring about drainage of the soil and compare favourably with sand drains. In fact Broms and Boman (1976) maintained that the drainage effect of one lime column was equivalent to that of two or three 100 mm wide band drains or three 150 mm diameter sand drains (see Chapter 4).

Lime piles can be installed in saturated clayey soils by means of a special metal tube with a closed tip (from 250 to 500 mm in diameter), holes being driven to depths of 5–8 m and spread at 1.5–2.5 m centres (Tsytovich *et al.*, 1971). Then the tube is withdrawn from the soil and the hole is filled with lumps of quicklime (Figure 10.13). Casing is used if the walls of the hole are unstable. The casing is extracted as the quicklime is placed. Lastly, the quicklime is packed by tamping. On contact with the groundwater, the quicklime gradually slakes, which means that the diameter of the pile may expand from 30 to 70% of its original size. Furthermore, a large amount of heat is released as slaking takes place, which causes some of the pore water in the surrounding soil to evaporate. Lime also penetrates the soil, thereby effecting its stabilization.

10.3 OTHER MATERIALS USED FOR SOIL STABILIZATION

Numerous other materials have been used for stabilizing soil. For example, pulverized fly ash (PFA) can be used by itself to improve the physical properties of a soil or in conjunction with lime or cement to form a binder. Pulverized fly ash is a pozzolan, that is, it reacts with CaO and water to form cementitious material. Because there is a slower gain in strength when PFA is mixed with lime, and because PFA has a greater sensitivity to low temperatures, cement has usually been preferred as the mixing agent. From practical experience, it has been found advantageous to use rapid-hardening cement. The early gain in strength provides a safeguard against the effects of adverse weather after compaction. Pulverized fly ash–cement or PFA–lime stabilization is best suited to sands and gravels with low clay contents. Well-graded gravel, for example, can be stabilized by the addition of 10% PFA (based on dry weight) and 5% ordinary Portland cement.

The compressive strength of the stabilized material is determined by the characteristics of the PFA, the cement, the degree of compaction and the efficiency of mixing. Generally PFA stabilizes readily with a reasonably low cement content. With some mix-in-place procedures, however, there is a risk that weak areas of poorly stabilized material will be formed if the cement content is too low. A minimum of 7% is advisable under such conditions. The premixed method, on the other hand, allows more effective control of cement and moisture contents.

Lime–pulverized fly ash slurry, injected under pressure has been used to stabilize embankments and levees in the United States (Baez *et al.*, 1992). Increases in strength of 15–30% have been recorded.

Bituminous material has been used to stabilize granular soils, providing cohesion and thereby enhancing their strength. In the case of cohesive soils, the addition of bitumen waterproofs the soil and in this way the loss of strength associated with increasing moisture content is reduced. The quantity of bitumen required for successful waterproofing of cohesive soils normally increases with increasing clay content. Even so, soils with similar clay contents may require different amounts of bitumen depending on their affinity for water. Like cement and lime stabilization, stabilization of organic soils with bitumen is not successful.

Bitumen may be applied hot or as cutbacks, or emulsions. Cutbacks, in which the bitumen is thinned by the addition of a volatile oil (e.g. paraffin or naphtha), are more often used than hot bitumen as the latter sometimes proves difficult to mix with soil. As the volatile evaporates from cutback bitumen, the bitumen itself is deposited in the soil. The higher the proportion of solvent in cutback bitumen, the lower is its viscosity and the greater is its volatility, giving a shorter curing time.

In temperate climates cutback bitumen can be used to stabilize sands since excess water can be removed by compaction. In the wet-sand mix, 4–10% of cutback bitumen is mixed with the wet sand to which 1–2% of hydrated lime has been added previously to assist coating. In the United Kingdom it has been found that once the silt fraction exceeds 5% the mixture becomes difficult to compact. In warmer climates 10% or more silt material can be tolerated, and in uniformly graded sands the figure may rise to 25%.

The amount of cutback bitumen, and hence the volatiles content, required for attaining maximum strength may be less than that required for most efficient mixing or compaction. This need for different volatile contents at different stages of construction can be allowed for by affording the stabilized soil a curing period during which volatiles evaporate. In other words, a period of time may be allowed between pulverization and mixing on the one hand and compaction on the other.

Bituminous emulsions are used cold and consist of a fine suspension of particles of bitumen in water. The bitumen is deposited as the suspension coagulates or breaks. Bituminous emulsions are generally only suitable for soil stabilization in climates where rapid drying conditions occur. Nevertheless, emulsions have been used in the United Kingdom, for example, to treat cohesive soils. In such instances cement or lime is added to the mix after treatment with bitumen. This causes the emulsion to break; it also absorbs some of the excess moisture as a result of hydration, and affords extra strength to the compacted soil. About 5–7.5% of emulsion and 3–5% cement or lime are added to soils treated in this way.

Construction methods used for bituminous stabilization are similar to those used for soil–cement mixtures. However, the optimum moisture content is usually below that necessary for efficient mixing. Accordingly, except for sands, it is often necessary to allow a period of time for the mix to dry before compaction takes place. In the mix-in-place method, bitumen is usually added in several passes, each layer being partially mixed before the next pass in order to avoid saturating the surface of the soil.

Some other soil stabilizers have a useful application in certain special circumstances, especially where temporary solutions are acceptable. Chemical additives that have been used for soil stabilization include phosphoric acid, sodium chloride, calcium chloride, sodium hydroxide, synthetic resins such as urea formaldehyde, and lignin.

The treatment with phosphoric acid is restricted to acid soils and is normally ineffective in sands and silts. This is because stabilization is brought about as a result of acid attack on clay minerals, dissolving their aluminium content. The latter is then precipitated as hydrated aluminium phosphate.

The addition of sodium chloride is effective in most types of soil, although it is not so useful in saline or highly organic soils. It aids compaction by prompting uniformity in the mix and reducing the amount of water loss due to evaporation. Thus higher densities are attained. However, because of its solubility it is easily leached from the soil by rain. Consequently, frequent applications are required unless the surface of the soil is sealed. Soils with high liquid limits usually respond well to the addition of sodium chloride.

The addition of calcium chloride to soil has similar effects to those of sodium chloride. However, it has an adverse influence on compaction and tends to increase the permeability of the soil.

Lateritic soils can be stabilized with the addition of sodium hydroxide, which is applied with the mix water and facilitates compaction. Sodium hydroxide reacts with aluminium-bearing minerals, notably kaolinite, and a substantial increase in strength occurs after an initial curing period. By contrast, the addition of sodium hydroxide to montmorillonitic soils proves detrimental in that it leads to a decrease in their strength.

The strength of sandy soils stabilized by urea formaldehyde depends on resin concentrations, soil characteristics (particularly particle size distribution and clay content), duration of gelling and pH value of the stabilized medium. In particular, the strength of such stabilized soils increases as the particle size distribution decreases, that is, as the permeability is reduced. The presence of montmorillonite tends to retard hardening and therefore lowers the strength achievable by such stabilization. In fact, the use of urea formaldehyde is recommended for stabilization of sand that has less than 3% clay fraction and a pH value less than 7.5. In such treatment the stabilizing liquid is pumped into holes in the soil, under pressures of up to 300 kN/m^2.

Lignin can be added to soil either in the form of sulphite liquor or as a powder. Unfortunately, however, lignin is soluble in water and consequently this type of soil treatment is not permanent. If sodium bichromate or potassium bichromate is added to sulphite liquor, then an insoluble gel results. Chrome–lignin has been used to stabilize soil, the chromium ion being an effective chelating agent, strengthening the bond between clay particles and lignin molecules. However, such treatment of soils is expensive and is therefore used only in special situations, for example, in stabilizing volcanic or chloritic soils.

10.4 THERMAL STABILIZATION

Clay soils harden on heating, and if heated to a high enough temperature they remain hard. This is due to the fact that changes occur in the crystalline structure of clay minerals above about 400°C, notably the loss of the (OH) group. These changes are irreversible and give rise to other significant changes in physical properties.

Basically, thermal stabilization consists of driving exhaust gases, from burning oil, at temperatures around 1000°C, into holes in the ground (Figure 10.14). Depths of treatment of up to 20 m have been obtained. In

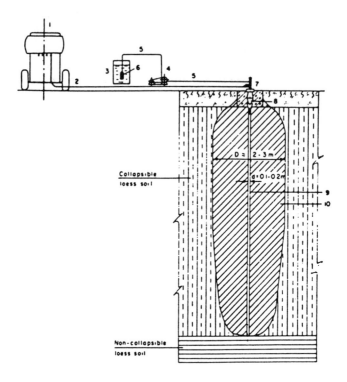

Figure 10.14 Scheme for deep thermal treatment of soil (after Litvinov, 1960): 1, compressor; 2, cold air pipeline; 3, liquid fuel container; 4, pressurized fuel pump; 5, fuel pump line; 6, filters; 7, nozzle; 8, combustion chamber; 9, borehole; 10, stabilized zone

the open-firing method two holes are bored in the soil so that they intersect. The combustion nozzle is placed over one and the combustion gases exit from the other. In the closed-firing method a single hole closed system is used in which the burner temperature is controlled by maintaining an excess air pressure. Air and gas pressure is typically 25–50% above atmospheric. Gas temperature, ranging from about 800°C for an air/fuel ratio of 3.5 to 2800°C for a ratio of 1 are achievable. Stabilization proceeds by progressively locating the flame deeper in the borehole. Heating is brought about by penetration into the pores of the soil of hot compressed air. Air temperatures must not exceed the fusion temperature of the soil. The latter method tends to be the more efficient, although it is not applicable to soils of low permeability. Many holes can be burned at the same time by linking a main fuel supply to a series of individually adjusted burner heads.

The distribution of temperature with depth in the soil depends on its porosity, moisture content and excess pore-water pressure, and the temperature of the gases injected. Until the free water enclosed in the pores of the soil is completely evaporated, the temperature does not exceed 100°C. Soils containing large proportions of organic colloids and colloidal minerals react at low temperatures. For instance, at 200–400°C they begin to develop some water resistance as fine particles begin to aggregate into granules. Further increase in temperatures leads to caking and ultimately to fusion. Thermal stabilization cannot be applied to soils that are saturated since the latent heat of evaporation of water makes it too expensive.

An alternative method of *in-situ* heating is to use electric heaters. In one system compressed air is blown through an electric heater at the top of a borehole. Temperatures from 500 to 1200°C can be achieved. In another, electric heaters are inserted in the borehole. Soil around the borehole is baked to form a column. Walls and structural mats can be formed by interlocking columns. Soils can be treated regardless of water content or gas permeability.

An appreciable reduction occurs in the plasticity index of a clay if heated to 400°C. Also the moisture absorption capacity of clay is appreciably reduced after it has been heated to 600°C. Initially the permeability of a clay soil increases on heating up to 600 or 700°C, above which, owing to the onset of fusion, it decreases slightly. The compressibility of clay soils is reduced by thermal treatment, as is their swelling capacity.

Thermal treatment has been used to prevent shear failure in clay soils and to stabilize clay slopes (Beles, 1957). It also has been used to stabilize collapsible loess by enhancing its strength appreciably (Litvinov, 1960). Burning oil or gas in a borehole 0.15–0.2 m in diameter stabilizes a column of loess some ten times as wide, and depths of 8–10 m can be stabilized in about 10 days. Loess is baked at a temperature somewhere between 300 and 1000°C (Evstatiev, 1988).

Appendix

Summary of soil-treatment methods

Method	Soils	Uses	Advantages	Disadvantages
Exclusion techniques				
1. Sheet piling	All types of soil except tills with numerous boulders	Almost unrestricted Cofferdams Trench and excavation support About 10 m depth	Plant readily available Rapid installation Steel piles can be incorporated in permanent works or recovered	Difficult to drive and maintain seal in soil with boulders. Vibration and noise of driving may not be acceptable Capital investment in piles can be high if re-usage is restricted Seal may not be perfect
2. Slurry trenches	Silts, sands and gravels	Almost unrestricted Extensive curtain walls round open excavations Depth, to about 20 m	A rapidly installed, cheaper form of diaphragm wall Can be keyed into impermeable strata such as clays or soft shales	Must be adequately supported Cost increases greatly with depth Costly to attempt to key into hard or irregular bedrock surfaces
3. Diaphragm walls (unreinforced and reinforced)	All soil types including those containing boulders (rotary percussion drilling suitable for penetrating rocks and boulders by reverse circulaton using bentonite slurry)	Deep basements Underground car parks Underground pumping stations Shafts Dry docks, etc. Depth, over 50 m	Can be designed to form part of a permanent foundation Particularly efficient for circular excavations Can be keyed into rock Minimum vibration and noise Treatment is permanent Can be used in restricted space Can be put down very close to existing foundations	High cost may make it uneconomical unless it can be incorporated into permanent structure There is an upper limit to the density of steel reinforcement that can be accepted

Appendix (continued)

Method	Soils	Uses	Advantages	Disadvantages
4. Contiguous-bored pile walls and secant piles	All soil types, but penetration through boulders may be difficult and costly	As for 3 Underpasses in stiff soils Significant depths can be treated	Can be used on small and confined sites Can be put down very close to existing foundations Minimum noise and vibration Treatment is permanent Secant piles provide impermeable barrier	Ensuring complete contact of all piles over their full length may be difficult in practice Joints may be sealed by grouting externally Efficiency of reinforcing steel not as high as for 3
5. Thin, grouted membranes	Silts and sands	As for 2	As for 2	The driving and extracting of the sheet-pile element used to form the membrane limits the depth achievable and the type of soil Also as for 2
Freezing				
6. Ammonia, freon/brine refrigeration	All types of soils with moisture content exceeding 10%	Formation of ice in the voids impedes water flow Shafts and deep narrow excavations	Imparts temporary mechanical strength to soils Treatment effective from working surface outwards Better for large applications of long duration	Treatment takes time to develop Initial installation costs are high and refrigeration plant is expensive Requires strict site control Some ground heave possible Not suitable if running water
7. Cryogenic liquids, e.g. liquid nitrogen	As for 6	As for 6 Frequently used as expedient	As for 6, but better for small applications of short duration or where quick freezing is required Can be used where water flow up to 50 m/day	Liquid nitrogen is expensive Requires strict site control Some ground heave possible
Drainage				
8. Filter drains	Sandy silts, silts, tills and clays	Site drainage, drainage of toes of embankments and slopes	Relatively low cost	Avoid fabric tearing otherwise filter can clog
9. Drainage galleries	Any water-bearing strata underlain by low permeability strata, suitable for tunnelling	Removal of large quantities of water for dam abutment, cut-offs, etc. Stabilization of slopes	Very large quantities of water can be drained into gallery and disposed of by conventional large-scale pumps	Very expensive Galleries may need to be concreted and grouted later
10. Sand and band drains	Soft clays, silts, organic soils	Consolidation of soils more rapidly over large areas to depths of 20 m	Easy to install, uniform result Low cost	Relatively slow
11. Lime columns	Silts, peat, clayey silts, clays	Drainage, consolidation and reinforcement of soil over large areas to depth of about 10m	Easy to install, rapid result	Non-uniformity High cost

Appendix (*continued*)

Method	Soils	Uses	Advantages	Disadvantages
Exclusion of groundwater by groundwater lowering				
12. Sump pumping	Clean gravels and coarse sands	Open, shallow excavations	Simplest pumping equipment	Fines easily removed from ground Encourages instability of formation
13. Wellpoint systems with suction pumps (including the machine-laid horizontal system)	Sandy gravels down to fine sands (with proper control can also be used in silty sands)	Open excavations including rolling-pipe trench excavations Horizontal system particularly pertinent for pipe trench excavations outside urban areas	Quick and easy to install in suitable soils Economical for short pumping periods of a few weeks	Difficult to install in open gravels or ground containing cobbles and boulders Pumping must be continuous and noise of pump may be a problem in a built-up area Suction lift is limited to about 4.0–5.5 m, depending on soils; if greater lowering is needed, multi-stage installation is necessary
14. Bored shallow wells with suction pumps	Sandy gravels to silty fine sands	Similar to wellpoint pumping More appropriate for installations to be pumped for several months or for use in silty soils where correct filtering is important	Generally costs less to run than a comparable wellpoint installation, so if pumping is required for several months costs should be compared Correct filtering can be controlled better than with well-points to prevent removal of fines from silty soils	Initial installation is fairly costly Pumping must be continuous and noise of pump may be a problem in a built-up area Suction is limited to about 4.0–5.5 m, depending on soils If greater lowering is needed, multi-stage installation is necessary
15. Deep-bored filter wells with electric submersible pumps (long-shaft pumps with motor mounted at wellhead used in some countries)	Gravels to silty fine sands	Deep excavation in, through or above water-bearing formations	No limitation on amount of drawdown as there is for suction pumping A well can be constructed to draw water from several layers throughout its depth Vacuum can be applied to assist drainage of fine soils Wells can be sited clear of working area No noise problem if mains electricity supply is available	High installation cost
16. Jet eductor system using high-pressure water to create vacuum as well as to lift the water	Sands (with proper control can also be used in silty sands and sandy silts)	Deep excavations in space so confined that multi-stage wellpointing cannot be used Usually more appropriate to low-permeability soils	No limitation on amount of drawdown Raking holes are possible	Initial installation is fairly costly Risk of flooding excavation if high-pressure watermain is ruptured Optimum operation difficult to control
17. Vacuum dewatering	Silts and silty clays	Vacuum pumps, sealed at surface	Rapid and uniform result	Not effective in cold regions

Appendix (*continued*)

Method	Soils	Uses	Advantages	Disadvantages
Electro-osmosis and electrochemical stabilization				
18. Electro-osmosis	Silts, silty clays and some organic soils	DC current causes water to flow from anode to cathode where it is removed Open excavation in appropriate soils or to speed dissipation of construction pore pressures	In appropriate soils can be used when no other water-lowering method is applicable	Installation and running costs are usually high Useless in saline soils Becomes less effective with time Rehydration occurs after treatment
19. Electrochemical stabilization	Normally consolidated silts, silty clays, clays, quick-clays and loess	Depth about 15 m, moderate areas Stabilizing chemicals moved into soil by electro-osmosis	Fairly fast Improve shear strength of soft clay without causing settlement (i.e. compressibility increased)	Installation and running costs are high Not useful in saline soils
Compaction techniques				
20. Preloading	Soft clays, silts organic soils, peat and loess	Requires earthmovers, fill and filter material Load applied sufficiently in advance of construction so that compression of soft soils is completed prior to development of site. Can use with drains	Easy and uniform result at low cost Reduced moisture content increased strength Large areas can be treated to significant depth	Very slow
21. Surcharge fills	Normally consolidated soft clays, silts, organic deposits, peat and loess	Earth moving equipment for loading sand or gravel for drainage blanket Fill in excess of that required permanently is applied to achieve a given amount of settlement in a shorter time, excess fill then removed Can use vertical drains to reduce consolidation time	Reduced water content, void ratio and compressibility, increased strength Faster than preloading without surcharge Moderate cost	
22. Prewetting (inundation)	Expansive and collapsing soils (e.g. loess)	Reduces differential movements	Very low cost Treats moderate areas over a depth of 4 m	Slow, uncontrollable
23. Hydrocompaction	Collapsible soils (e.g. silts in arid areas and loess)	To avoid settlement in collapsible soils	Rapid and low cost	Applies only to loose, arid silts and loess

Appendix (*continued*)

Method	Soils	Uses	Advantages	Disadvantages
24. Vibrocompaction (a) Vibroflotation	Gravels and sands	Densify soil with vibroflot suspended from crane to depths of 20 m or so over site	Form uniform ground conditions at moderate cost	Relative slow procedure
(b) Vibro-replacement	Silts and clays with strengths above around 20 kN/m^2	Strengthening and densification of ground by inclusion of stone columns which can act in similar fashion to piles	Significant improvement of bearing capacity and overall reduction of settlement	Ground around columns may settle
(c) Terra-probe	Saturated or dry clean sand	Vibratory pile driver and 750 mm dia. open steel casing. Treat to a depth 20 m (ineffective above 3–4 m depth) Densification by vibration: liquefaction induced settlement under overburden	Drainage may be facilitated by columns Can obtain relative densities of up to 80%. Ineffective in some sands Rapid, simple, good underwater Moderate cost	Soft underlayers may damp vibrations, difficult to penetrate stiff overlayers, not good in partly saturated soils
25. Dynamic compaction	Most types of soils	Heavy tamper (up to 200 t) suspended from crane impacts ground in a number of passes	Can treat large areas to depth of 30 m or so. Bearing capacity greatly enhanced	In cohesive soils must wait for dissipation of pore-water pressure between passes
26. Compaction piles	Loose sandy soils, partly saturated clayey soils, loess peat	Densification by displacement pile by vibration during driving Treat to depth of up to 20 m	Can obtain high densities, good uniformity Useful in soils with fines, uniform compaction, easy to check results	Slow limited improvement in upper 1–2 m Moderate to high cost
27. Electric shock	Wet sands and silts	High volt impulse generator and condenser required Treats depth up to 10 m	Simple Low cost	Not useful above 1 m depth
28. Blasting	Partially saturated and saturated sands, silts loess	Explosives, drilling gear necessary to treat large areas to a depth of 30 m Shock waves and vibrations cause liquefaction, displacement and settlement	Rapid, cheap can attain relative densities between 70 and 80%	Variable properties in partially saturated soils. No improvement near surface
29. Mechanical compaction	Gravels, sands, silts and clays	Various types of rollers used depending on soil type	Rapid, simple for thin layers used at low cost	

Appendix (continued)

Method	Soils	Uses	Advantages	Disadvantages
Reinforcement				
30. Reinforced earth	Sands, silts and gravels	Geotextiles, meshes or metal strips used to enhance strength of soil in embankments, retaining walls, etc.	Strong, simple to place Low to moderate cost	
31. Soil nailing, and root piles	Most types of soils	Treatment of slopes and foundations Root piles also used for underpinning	Enhances strength of ground at moderate cost	
Grouting				
32. Cement, clay and cement–clay grouts	Sands and gravels	Filling voids to exclude water To form relatively impermeable barriers (vertical or horizontal) Suitable for conditions where long-term flexibility is desirable, e.g. cores of dams Treatment can go to great depth	Equipment is simple and can be used in confined spaces Treatment is permanent Grout is introduced by means of a sleeved grout pipe which limits its spread Can be sealed to an irregular or hard stratum Lowest costs of grouting	A comparatively thick barrier is needed to ensure continuity At least 4 m of natural cover needed (or equivalent)
33. Silicate grouts	Medium and coarse sands and gravels, loess	As for 32 but non-flexible	Comparatively high mechanical strength High degree of control of grout spread Simple means of injection by lances Indefinite life Favoured for underpinning works below water level	Comparatively high cost of chemicals Requires at least 2 m of natural cover or equivalent Treatment can be incomplete in silty material or in presence of silt or clay lenses
34. Resin grouts	Silty fine sands	As for 32 but only some flexibility	Can be used in conjunction with clay–cement grouts for treating finer strata	High cost so usually economical only on larger civil engineering works Requires strict site control
35. Compaction grouting	Soft, fine-grained soils, foundation soils with large voids, loess	Highly viscous grout acts as radial hydraulic jack when pumped in under high pressure Depth of treatment usually a few metres	Grout bulbs formed within compressed soil matrix Good for correction of differential settlements, filling large voids	Careful control required Relatively high cost
36. Jet grouting	Most types of soils	High-speed jets at depth excavate, inject or mix stabilizer with soil to form columns or panels	Solidified columns and walls Useful in soils that cannot be permeation grouted	Precision in locating treated zones

Appendix (*continued*)

Method	Soils	Uses	Advantages	Disadvantages
Soil stabilization				
37. Cement and lime stabilization, other additives (e.g. bitumen, PFA, various chemicals)	Most types of soils	Soil and additive mixed either on site or in place and compacted in layers Large areas can be treated usually to shallow depth Road and runway construction canal linings Cement cushions	Enhanced engineering performance, uniform ground conditions Rapid and simple at low to moderate cost	Additives not equally effective in all soils
38. Mix-in-place piles, etc.	Silts and clays (including soft and expansive types), loess	Lime network formed by intrusion into cracks Treat moderate to large areas to depths of 10–20 m or so	High strength, rapid effect, at moderate cost	
39. Heating	Unsaturated silts, clays and loess	Small areas treated to relatively shallow depth by baking Stabilization of slopes and foundations	Rapid and permanent strength gain	Not useful in saturated soils Moderate to high cost

References

Ahnberg, H., Bengtsson, P. E. and Holm, G. (1989) Prediction of strength of lime columns. *Proc. 12th International Conf. on Soil Mechanics and Foundation Engineering, Rio de Janeiro*, Vol. 2, pp. 1327–30.

Aldrich, H. P. (1965) Precompression for support of shallow foundations. *Proc. American Society Civil Engineers, Journal Soil Mechanics and Foundations Division*, **91** (SM2), 5–20.

Al-Rawi, N. M. and Awad, A. A. A. (1981) Permeability of lime stabilized soils. *Proc. American Society of Civil Engineers, Journal Transportation Engineering Division*, **107** (TEI), 25–35.

Andrawes, K. Z., McGown, A., Mashour, M. M. and Wilson-Fahmy, R. F. (1980) Tension resistant inclusions in soils. *Proc. American Society Civil Engineers, Journal Geotechnical Engineering Division*, **106** (GT12), 1313–26.

Anon. (1955) Drainage and erosion control. Subsurface drainage for airfields. *Engineering Manual, Military Construction*, Chapter 2, United States Army Corps of Engineers, Washington, D.C.

Anon. (1957) Chemical grouting. Report of the Task Committee on Chemical Grouting. *Proc. American Society of Civil Engineers, Journal Soil Mechanics and Foundation Division*, **83** (SM4), 1–101.

Anon. (1962) Cement grouting. Progress Report of Task Committee on Cement Grouting. *Proc. American Society Civil Engineers, Journal Soil Mechanics and Foundations Division*, **88** (SM2), 49–98.

Anon. (1975) *Earth Manual*, US Bureau of Reclamation, Washington, D.C.

Anon. (1976) *Methods of Tests for Soils for Civil Engineering Purposes*, BS 1377, British Standards Institution, London.

Anon. (1981a) *Code of Practice on Earthworks*, BS 6031, British Standards Institution, London.

Anon. (1981b) *Code of Practice on Site Investigation*, BS 5930, British Standards Institution, London.

Anon. (1982) *Lime Stabilization Construction Manual*, 7th edn, National Lime Association, Washington, D.C.

Anon. (1983) *Lining of Waste Impoundment and Disposal Facilities*, Office of Solid Waste and Emergency Resources, Publication No. SW870, US Environmental Protection Agency, Washington, D.C.

Anon. (1984) *A Geotextile Design Guide*, Low Bros, Dundee.

Anon. (1985a) First 'Stent Wall' installed at Kingston upon Thames. *Ground Engineering*, **18** (7), 27–31.

Anon. (1985b) Specifications for cast-in-place diaphragm walling. *Ground Engineering*, **18** (6), 11–19.

Anon. (1986a) *Specification for Highway Works*, Department of Transport, HMSO, London.

Anon. (1986b) Design standards No. 13: Embankment dams. *Soil-Cement Slope Protection*, US Bureau of Reclamation, Denver, Chapter 17.

Anon. (1988) Diaphragm walling for Sizewell B sets records. *Ground Engineering*, **21** (3), 19–25.

Anon. (1989a) Dewatering myths. *Ground Engineering*, **22** (2), 12–15.

Anon. (1989b) The use of geotextiles and geocomposites in the United Kingdom. *Proc. American Society of Civil Engineers, Journal Construction Engineering and Management Division*, **115** (2), 258–69.

Anon. (1990a) Tropical residual soils, Engineering Group Working Party Report. *Quarterly Journal of Engineering Geology*, **23**, 1–101.

Anon. (1990b) Geotextile groundwork. *Ground Engineering*, **23** (3), 14–22.

Anon. (1990c) Cement grouts. *Ground Engineering*, **23** (7), 18–21.

Anon. (1990d) State-of-the-art report on soil-cement. *American Concrete Institute Materials Journal*, **87** (4), 395–417.

Anon. (1991) Jet set style. *Ground Engineering*, **28** (8), 22–3.

Arcones, A., Ruiz de Temiño, R. and Soriano, A. (1985) Strengthening a structural fill by claquage. *Proc. 11th International Conf. on Soil Mechanics and Foundation Engineering, San Francisco*, Vol. 3, pp. 1677–82.

Arenzana, L., Krizek, R. J. and Pepper, S. F. (1989) Injection of dilute microfine cement suspensions into fine sands. *Proc. 12th International Conf. on Soil Mechanics and Foundation Engineering, Rio de Janiero*, Vol. 2, pp. 1331–4.

Auld, F. A. (1985) Freeze wall strength and stability design problems in shaft sinking. *Proc. 4th International Symp. on Ground Freezing, Sapporo*, (eds S. Kinosita and M. Fukuda), Balkema, Rotterdam, pp. 345–9.

Baez, J. I., Borden, R. H. and Henry, J. H. (1992) Rehabilitation of lower Chariton river levee by lime/flyash slurry injection, in *Transportation Research Record*, Transportation Research Board, National Research Council, Washington, D.C., pp. 117–25.

Barker, D. H. (1986) Biodegradable geotextiles for erosion control and slope stabilization. *Civil Engineering*, **81** (June), 13–15.

Barron, R. A. (1948) Consolidation of fine grained soils by drain wells. *Transactions American Society Civil Engineers*, **133**, 718–54.

Baumann, V. and Bauer, G. E. A. (1974) The performance of foundations on various soils stabilized by the vibrocompaction method. *Canadian Geotechnical Journal*, **11**, 509–30.

Beles, A. A. (1957) Le traitment thermique du sol. *Proc. 4th International Conf. on Soil Mechanics and Foundation Engineering, Paris*, Vol. 3, pp. 266–9.

Bell, F. G. (1975) *Methods of Treatment of Unstable Ground*, Newnes-Butterworths, London.

Bell, F. G. (1976) The influence of the mineral content of clays on their stabilization with cement. *Bulletin Association of Engineering Geologists*, **13**, 267–78.

Bell, F. G. (1988) Lime stabilization of clay soils: Part 1, Basic principles. *Ground Engineering*, **21** (1), 10–15.

Bell, F. G. and Cashman, P. M. (1986) Groundwater control by groundwater lowering, in *Groundwater in Engineering Geology*, Engineering Geology Special

Publication No. 3 (eds J. C. Cripps, F. G. Bell and M. G. Culshaw), The Geological Society, London, pp. 471–86.

Bell, F. G. and Mitchell, J. K. (1986) Control of groundwater by exclusion, in *Groundwater in Engineering Geology*, Engineering Geology Special Publication No. 3 (eds J. C. Cripps, F. G. Bell and M. G. Culshaw), The Geological Society, London, pp. 429–43.

Bell, F. G. and Coulthard, J. M. (1990) Stabilization of glacial deposits of the Middlesbrough Area with cementitious material. *Proc. 6th International Congress, International Association of Engineering Geology, Amsterdam* (ed. D. G. Price), Balkema, Rotterdam, Vol. 3, pp. 1797–807.

Bergado, D. T., Asakami, H. and Alfaro, M. C. (1991) Smear effects of vertical drains on soft Bangkok Clay. *Proc. American Society Civil Engineers, Journal Geotechnical Engineering Division*, **117** (10), 1509–30.

Bertram, G. E. (1940) *An Experimental Investigation of Protective Filters*, Harvard University Publication No. 267.

Bevan, D. J. and Johnson, D. (1989) The use of geogrid to enable treatment of peaty soils by vibrodisplacement. *Ground Engineering*, **22** (2), 34–5.

Bjerrum, L. (1972) Embankments on soft ground. General report. *Proc. American Society Civil Engineers, Speciality Conf. on Performance of Earth and Earth Supported Systems*, Vol. 2, pp. 1–54.

Bjerrum, L., Moun, J. and Eide, O. (1967) Application of electro-osmosis to a foundation problem in a Norwegian quick clay. *Geotechnique*, **17**, 214–35.

Bodocsi, A. and Bowers, M. T. (1991) Permeability of acrylate, urethane and silicate grouted sands with chemicals. *Proc. American Society Civil Engineers, Journal Geotechnical Engineering Division*, **117** (8), 1227–44.

Boyes, R. G. H. (1975) The versatility of diaphragm walling. *Civil Engineering*, **70** (December), 57–9.

Brand, E. W. and Pang, P. L. R. (1991) Durability of geotextiles to outdoor exposure in Hong Kong. *Proc. American Society Civil Engineers, Journal Geotechnical Engineering Division*, **117** (7), 979–1000.

Brandl, H. (1981) Alteration of soil parameters by stabilization with lime. *Proc. 10th International Conf. on Soil Mechanics and Foundation Engineering, Stockholm*, Vol. 3, pp. 587–94.

Braun, B., Shuster, J. A. and Burnham, E. W. (1979) Ground freezing for support of open excavations. *Proc. 1st Symp. on Ground Freezing, Bochum* (ed. H. L. Jessberger), Elsevier, Amsterdam, pp. 429–53; also *Engineering Geology*, **13**.

Broms, B. B. and Boman, P. (1976) Stabilization of deep cuts with lime columns. *Proc. 6th European Conf. on Soil Mechanics and Foundation Engineering, Vienna*, Vol. 1, pp. 207–10.

Broms, B. B. and Boman, P. (1979) Lime columns. A new foundation method. *Proc. American Society Civil Engineers, Journal Geotechnical Engineering Division*, **105** (GT4), 539–56.

Broms, B. B. and Shirlaw, J. N. (1987) Stabilization of waste ponds at Changi depot. *Proc. 9th South East Asian Geotechnical Conf., Bangkok*, pp. 8/13–8/24.

Brons and De Kruijff, H. (1985) The performance of sand compaction piles. *Proc. 11th International Conf. on Soil Mechanics and Foundation Engineering, San Francisco*, Vol. 3, pp. 1638–86.

Brown, R. E. (1977) Vibroflotation compaction of cohesionless soils. *Proc. American Society Civil Engineers, Journal Geotechnical Engineering Division*, **103** (GT12), 1437–51.

Bridle, R. J. (1989) Soil nailing-analysis and design. *Ground Engineering*, **22** (6), 52–6.

Bridle, R. J. and Barr, B. I. G. (1990) Soil nailing. Discussion. *Ground Engineering*, **23** (5), 30–1.

Bruce, D. A. and Jewell, R. A. (1986) Soil nailing: application and practice – Part 1. *Ground Engineering*, **19** (8), 10–15.

Bruce, D. A. and Jewell, R. A. (1987) Soil nailing: application and practice – Part 2. *Ground Engineering*, **20** (1), 21–33.

Burwell, E. B. (1958) Cement and clay grouting of foundations. Practice of the Corps of Engineers. *Proc. American Society of Civil Engineers, Journal Soil Mechanics and Foundation Division*, **84** (Paper 1551), 1551/1–1551/22.

Cambefort, H. (1967) *Injection des Sols*, Eyrolles, Paris.

Cambefort, H. (1977) Principles and applications of grouting. *Quarterly Journal of Engineering Geology*, **10**, 57–96.

Cannon, E. W. (1976) Fabrics in civil engineering. *Civil Engineering*, **71** (March), 39–42.

Cantoni, R., Collatta, T., Ghionna, V. N. and Moretti, P. C. (1989) A design method for reticulated micropile structures in sliding slopes. *Ground Engineering*, **22** (4), 41–6.

Caron, C. (1965) Etude Physico-Chimique des Gels de Silice. *Annals de l'Institut du Batiment et des Travaux Public*, No. 207–8.

Caron, C., Cattlin, P. and Herbst, T. F. (1975) Injections, in *Foundation Engineering Handbook* (eds H. F. Winterkorn and H. F. Fang), Van Nostrand Reinhold, New York, pp. 337–53.

Carpentier, R., De Wolf, P., Van Damme, L., De Rouck, J. and Bernard, A. (1985) Compaction blasting in offshore harbour construction. *Proc. 11th International Conf. on Soil Mechanics and Foundation Engineering*, San Francisco, Vol. 3, pp. 1687–92.

Casagrande, A. (1948) Classification and identification of soils. *Transactions American Society Civil Engineers*, **113**, 901–92.

Casagrande, L. (1952) Electro-osmosis stabilization of soils. *Journal Boston Society Civil Engineers*, **39**, 51–83.

Cashman, P. M. (1975) Control of groundwater by groundwater lowering, in *Methods of Treatment of Unstable Ground* (ed. F. G. Bell), Newnes-Butterworths, London, pp. 12–25.

Cashman, P. M. and Haws, E. T. (1970) Control of groundwater by groundwater lowering, in *Ground Engineering* (ed. J. S. David), Institution of Civil Engineers, London, pp. 23–32.

Cedergren, H. (1986) *Seepage, Drainage and Flow Nets*, 2nd edn, Wiley, New York.

Chi, B. P. C. and Yang, Jiann-Shi (1985) Chemical grouting in Taipei Basin. *Proc. 11th International Conf. on Soil Mechanics and Foundation Engineering*, San Francisco, Vol. 3, pp. 1693–6.

Choa, V. (1989) Drains and vacuum preloading pilot test. *Proc. 12th International Conf. on Soil Mechanics and Foundation Engineering*, Rio de Janeiro, Vol. 2, pp. 1347–50.

Chung, Y. T., Chung, S. T. and Wu, W. K. (1987) Improvements in hydraulic sandy fills by compaction piles. *Proc. 9th South East Asian Geotechnical Conf.*, Bangkok, pp. 8/57–8/68.

Clayton, C. R., Simons, N. E. and Matthews, M. C. (1982) *Site Investigation*, Granada, London.

Clevenger, W. A. (1958) Experiences with loess as a foundation material. *Transactions American Society Civil Engineers*, **123**, 151–80.

Cooke, T. F. and Rebenfeld, L. (1988) Effect of chemical composition on physical structure of geotextiles on their durability. *Geotextiles and Geomembranes*, **7**, 7–22.

Coomber, D. B. (1986) Groundwater control by jet grouting, in *Groundwater in Engineering Geology,* (eds J. C. Cripps, F. G. Bell and M. G. Culshaw), Engineering Geology Special Publication No. 3, The Geological Society, London, pp. 445–54.

Croft, J. B. (1968) The problem of predicting the suitability of soils for cementitious stabilization. *Engineering Geology,* **2**, 397–424.

D'Appolonia, E. D. (1953) *Loose Sands – Their Compaction by Vibroflotation.* Special Technical Publication 156, American Society for Testing Materials, Philadelphia, pp. 138–54.

D'Appolonia, D. J., Whitman, R. V. and D'Appolonia, E. D. (1969) Sand compaction by vibratory rollers. *Proc. American Society Civil Engineers, Journal Soil Mechanics and Foundations Division,* **95**, 263–83.

Dastidor, A. G., Gupta, S. and Gosh, T. K. (1969) Application of sandwicks in a housing project. *Proc. 7th International Conf. on Soil Mechanics and Foundation Engineering, Mexico City,* Vol. 2, pp. 89–94.

De Paoli, B., Mascardi, C. and Stella, C. (1989) Construction and quality control of a 100 m deep diaphragm wall. *Proc. 12th International Conf. on Soil Mechanics and Foundation Engineering, Rio de Janeiro,* Vol. 3, pp. 1479–82.

Dugnani, G., Guatteri, G., Roberti, P. and Mosiici, P. (1989) Sub-horizontal jet grouting applied to a large urban twin tunnel in Campinas, Brazil. *Proc. 12th International Conf. on Soil Mechanics and Foundation Engineering, Rio de Janeiro,* Vol. 2, pp. 1351–4.

Dupas, J. M. and Pecker, A. (1977) Static and dynamic properties of soil-cement. *Proc. American Society of Civil Engineers, Journal Geotechnical Engineering Division,* **105** (GT3), 413–36.

Elias, V. (1990) *Durability/Corrosion of Soil Reinforced Structures.* Federal Highway Administration, Publication No. FHWA-RD-89-186, Department of Transportation, McLean, Virginia.

Elias, V. and Juran, I. (1991) *Soil Nailing for Stabilization of Highway Slopes and Excavations,* Federal Highway Administration, Report No. FHWA-RD-89-198, Department of Transportation, McLean, Virginia.

Evstatiev, D. (1988) Loess improvement methods. *Engineering Geology,* **25**, 341–66.

Ferworn, D. E. and Weatherby, D. E. (1992) *A Contractor's Experience with Soil Nailing,* Transportation Research Record, National Research Council, Washington, D.C.

Floess, C. H., Lacy, H. S. and Gerken, D. E. (1989) Artificially frozen ground tunnel: a case history. *Proc. 12th International Conf. on Soil Mechanics and Foundation Engineering, Rio de Janeiro,* Vol. 2, pp. 1445–8.

Fluet, J. E. (1988) Geosynthetics for soil improvement: a general report and keynote address, in *Geosynthetics for Soil Improvement* (ed. R. D. Holtz), Geotechnical Special Publication No. 18, American Society Civil Engineers, New York, pp. 1–21.

Frivik, P. E. and Thorbergsen, E. (1981) Thermal design of artificial soil freezing systems. *Proc. 2nd International Symp. on Ground Freezing, Trondheim* (eds P. E. Frivik, H. Janbu, R. Saetersdal and L. I. Finorud); *Engineering Geology,* Vol. 28 (Special Issue), pp. 189–201.

Fukuoka, M. (1986) Fabric retaining wall with multiple anchors. *Proc. 3rd International Conf. on Geotextiles, Vienna,* Vol. 2, pp. 435–40.

Fukuoka, M. and Imamura, Y. (1982) Fabric retaining walls. *Proc. 2nd International Conf. on Geotextiles, Las Vegas,* Vol. 2, pp. 575–80.

Gallavresi, F. (1985) Ground freezing-application of mixed method (brine and liquid nitrogen). *Proc. 4th International Symp. on Ground Freezing, Sapporo* (eds S. Kinosita and M. Fukuda), Balkema, Rotterdam, Vol. 1, pp. 928–39.

Gambin, M. P. (1987) Dynamic consolidation of volcanic ash in Sumatra. *Proc. 9th South East Asian Geotechnical Conf., Bangkok*, pp. 8/145–8/158.

Gassler, G. (1992) *German Practice of Soil Nailing*, Transportation Research Record, Transportation Research Board, National Research Council, Washington, D.C.

Gassler, G. and Gudehus, G. (1981) Soil nailing: some aspects of a new technique. *Proc. 10th International Conf. on Soil Mechanics and Foundation Engineering, Stockholm*, Vol. 3, pp. 665–70.

Gerber, A. and Harmse, H. J. von M. (1987) Proposed procedure for the identification of dispersive soils by chemical testing. *The Civil Engineer in South Africa*, **29**, 397–9.

Giroud, J. P. (1982) Filter criteria for geotextiles. *Proc. 2nd International Conf. on Geotextiles, Las Vegas*, Vol. 1, pp. 103–8.

Giroud, J. P. (1986) From geotextiles to geosynthetics: a revolution in geotechnical engineering. *Proc. 3rd International Conf. on Geotextiles, Vienna*, Vol. 1, pp. 1–18.

Glossop, R. (1968) The rise of geotechnology and its influence on engineering practice. *Geotechnique*, **18**, 107–17.

Gonze, P., Lejeune, M., Thimas, J. F. and Monjoie, A. (1985) Sand ground freezing for the construction of a subway station in Brussels. *Proc. 4th International Symp. on Ground Freezing, Sapporo* (eds S. Kinosita and M. Fukuda), Balkema, Rotterdam, Vol. 1, pp. 277–83.

Goto, S. and Iguro, M. (1989) The world's first high strength, super deep slurry wall. *Proc. 12th International Conf. on Soil Mechanics and Foundation Engineering, Rio de Janeiro*, **3**, 1487–90.

Greenwood, D. A. (1970) Mechanical improvement of soils below ground surface. *Proc. Symp. on Ground Engineering, Institution of Civil Engineers*, Thomas Telford Press, London, 11–21.

Gudehus, G. and Schwarz, W. (1985) Stabilization of creeping slopes by dowels. *Proc. 11th International Conf. on Soil Mechanics and Foundation Engineering, San Francisco*, Vol. 4, pp. 1697–1700.

Hansbo, S. (1978) Dynamic consolidation of soil by a falling weight. *Ground Engineering*, **11** (5), 27–30.

Hansbo, S. (1979) Consolidation of clay by band-shaped prefabricated drains. *Ground Engineering*, **12** (5), 16–25.

Hansbo, S. (1987) Design aspects of vertical drains and lime column installations. *Proc. 9th South East Asian Geotechnical Conf., Bangkok*, Balkema, Rotterdam, pp. 8/1–8/12.

Hansmire, W. H., Russell, H. A., Rawnsley, R. P. and Abbott, E. L. (1989) Field performance of structural slurry wall. *Proc. American Society Civil Engineers, Journal Geotechnical Engineering Division*, **115** (2), 141–56.

Harris, J. S. (1989) State-of-the-art: tunnelling using artificial frozen ground. *Proc. 5th International Symp. on Ground Freezing, Nottingham* (eds R. H. Jones and J. T. Holden), Balkema, Rotterdam, Vol. 1, pp. 245–53.

Hilf, J. W. (1975) Compacted fill, in *Foundation Engineering Handbook* (eds H. F. Winterkorn and H. Y. Fang), Van Nostrand Reinhold, New York, pp. 304–11.

Hoare, D. J. (1978) Permeable synthetic fabric membranes. II – Factors affecting their choice and control in geotechnics. *Ground Engineering*, **11** (8), 25–31.

Hoare, D. J. (1987) Geotextiles, in *Ground Engineer's Reference Book* (ed. F. G. Bell), Butterworths, London, pp. 34/1–34/18.

Hobbs, N. B. (1986) Mire morphology and properties and behaviour of some British and foreign peats. *Quarterly Journal Engineering Geology*, **19**, 7–80.

Holm, G., Bredenberg, H. and Broms, B. (1981) Lime columns as foundation for light structures. *Proc. 10th International Conf. on Soil Mechanics and Foundation Engineering, Stockholm*, Vol. 3, pp. 687–94.

Honjo, Y. and Veneziano, D. (1989) Improved filter criterion for cohesionless soils. *Proc. American Society Civil Engineers, Journal Geotechnical Engineering Division*, **115** (1), 75–96.

Hughes, J. M. O. and Withers, N. J. (1974) Reinforcing soft cohesive soil with stone columns. *Ground Engineering*, **7** (3), 42–9.

Hutchinson, M. T., Daw, G. P., Shotton, P. G. and James, A. N. (1975) The properties of bentonite slurries used in diaphragm walling and their control. *Proc. Conf. on Diaphragm Walls and Anchorages, London*, Institution of Civil Engineers, pp. 33–40.

Ingles, O. G. and Metcalf, J. B. (1972) *Soil Stabilization*, Butterworths, Sydney.

Ingold, T. S. (1984a) Geotextiles as filters. *Ground Engineering*, **17** (2), 29–44.

Ingold, T. S. (1984b) Geotextiles as earth reinforcement in the United Kingdom. *Ground Engineering*, **17** (3), 29–32.

Ingold, T. S. (1991) Partial factor design of polymer reinforced soil walls. *Ground Engineering*, **24** (5), 34–8.

Ischy, E. and Glossop, R. (1962) An introduction to alluvial grouting. *Proc. Institution of Civil Engineers*, **22**, 449–74.

Jessberger, H. L. (1985) The application of ground freezing to soil improvement in engineering practice, in *Recent Developments in Ground Improvement Techniques* (eds A. S. Balasubramanian, S. Chandra, D. T. Bergado, J. S. Younger and F. Prinzl), Balkema, Rotterdam, pp. 469–82.

Jewell, R. A. and Pedley, M. J. (1990a) Soil nailing design: the role of bending stiffness. *Ground Engineering*, **23** (2), 30–6.

Jewell, R. A. and Pedley, M. J. (1990b) Soil nailing design: the role of bending stiffness. Discussion. *Ground Engineering*, **23** (5), 32–3.

Jewell, R. A., Milligan, G. W. E., Sarsby, R. W. and Dubois, D. (1984a) Interaction between soil and geogrids. *Proc. Symp. on Polymer Grid Reinforcement in Civil Engineering, London*, Institution of Civil Engineers, Thomas Telford Press, London, Paper 1.3.

Jewell, R. A., Paine, N. and Woods, R. I. (1984b) Design methods for steep reinforced embankments. *Proc. Symp. on Polymer Grid Reinforcement in Civil Engineering, London*, Institution of Civil Engineers, Thomas Telford Press, London, pp. 70–81.

Johnson, S. J. (1958) Cement and clay grouting of foundations: grouting with clay–cement grouts. *Proc. American Society of Civil Engineers, Journal Soil Mechanics and Foundations Division*, **84** (SM1, Paper 1545), 1545/1–1545/12.

Johnson, S. J. (1970) Precompression for improving foundation soils. *Proc. American Society Civil Engineers, Journal Soil Mechanics and Foundations*, **96** (SM1), 145–75.

Jones, J. S. (1981) Engineering practice in artificial ground freezing. State-of-the-art report. *Engineering Geology* (Special Issue), **18**, 313–26.

Jones, C. J. F. P. (1985) *Reinforced Earth*, Butterworths, London.

Jones, C. J. F. P., Murray, R. T., Temporal, J. and Mair, R. J. (1985) First application of anchored earth. *Proc. 11th International Conf. Soil Mechanics and Foundation Engineering, San Francisco*, Vol. 3, pp. 1709–12.

Joshi, R. C., Natt, G. S. and Wright, P. J. (1981) Soil improvement by lime-fly ash slurry injection. *Proc. 10th International Conf. on Soil Mechanics and Foundation Engineering, Stockholm*, Vol. 3, pp. 707–12.

Juran, I., Ider, H. M. and Farrag, K. (1990a) Strain compatibility analysis for geosynthetic reinforced soils walls. *Proc. American Society of Civil Engineers, Journal Geotechnical Engineering Division*, **116** (2), 303–29.

Juran, I., Baudrand, G., Farrag, K. and Elias, V. (1990b) Design of soil nailed retaining structures, in *Design and Performance of Earth Retaining Structures* (eds P. C. Lambe and L. A. Hansen), American Society Civil Engineers, Geotechnical Special Publication No. 25, New York, pp. 644–59.

Kamauchi, H. and Kitamori, I. (1985) Improvement of soft ground bearing capacity using synthetic meshes. *Journal of Geotextiles and Geomembranes*, **2**, 3–32.

Karol, R. H. (1968) Chemical grouting technology. *Proc. American Society Civil Engineers, Journal Soil Mechanics and Foundations Division*, **94** (SM1), 175–204.

Karol, R. H. (1982) Chemical grouts and their properties, in *Grouting in Geotechnical Engineering, Speciality Conf., New Orleans* (ed. W. H. Baker), American Society Civil Engineers, New York, pp. 359–77.

King, J. C. and Bush, E. G. W. (1963) Grouting of granular materials. *Symp. on Grouting, Transactions American Society of Civil Engineers*, **128**, 1279–1310.

Klein, A. M. and Polivka, M. (1958) Cement and clay grouting of foundations: on the use of admixtures in cement grouts. *Proc. American Society of Civil Engineers, Journal Soil Mechanics and Foundations Divisions*, **84** (SM1, Paper 1547), 1547/1–1547/24.

Klein, J. (1989) State-of-the-art: engineering design of shafts. *Proc. 5th International Symp. on Ground Freezing, Nottingham* (eds R. H. Jones and J. T. Holden), Balkema, Rotterdam, Vol. 1l, pp. 235–44.

Koerner, R. M. and Robins, J. C. (1986) In situ stabilization of slopes using nailed geosynthetics. *Proc. 3rd International Conf. on Geotextiles, Vienna*, Vol. 2, pp. 345–400.

Kravetz, G. A. (1958) Cement and clay grouting of foundations: the use of clay in pressure grouting. *Proc. American Society of Civil Engineers, Journal Soil Mechanics and Foundation Division*, **84** (SM1, Paper 1546), 1546/1–1546/30.

Kujala, K. (1989) Frost action and the mechanical properties of an artificially frozen test plot. *Proc. 12th International Conf. on Soil Mechanics and Foundation Engineering, Rio de Janeiro*, Vol. 2, pp. 1449–54.

Kujala, K., Halkola, H. and Lahtinen, P. (1985) Design parameters for deep stabilized soil evaluated from in situ and laboratory tests. *Proc. 11th International Conf. on Soil Mechanics and Foundation Engineering, San Francisco*, Vol. 3, pp. 1717–20.

Kuno, G., Kutara, K. and Miki, H. (1989) Chemical stabilization of soft soils containing humic acid. *Proc. 12th International Conf. on Soil Mechanics and Foundation Engineering, Rio de Janeiro*, Vol. 2, pp. 1381–4.

Lawson, C. R. and Ingles, O. G. (1982) Long term performance of MESL road sections in Australia. *Proc. 2nd International Conf. on Geotextiles, Las Vegas*, Vol. 2, pp. 535–9.

Lee, S. L., Karunaratne, G. P., Yong, K. Y. and Ramaswamy, S. D. (1989) Performance of fibre drain in consolidation of soft soils. *Proc. 12th International Conf. on Soil Mechanics and Foundation Engineering, Rio de Janeiro*, Vol. 3, pp. 1667–70.

Leflaive, E. (1988) Texsol: already more than 50 successful applications. *Proc. International Symp. on Theory and Practice of Earth Reinforcement, Tokyo*, pp. 541–5.

Leshchinsky, D. and Boedeker, R. H. (1989) Geosynthetic reinforced soil structures. *Proc. American Society of Civil Engineers, Journal Geotechnical Engineering Division*, **115** (10), 1459–78.

Lewis, W. A. and Parsons, A. W. (1958) *An Investigation of the Performance of a 3¾ ton Vibrating Roller for Compacting Soil*, DSIR Road Research Laboratory, Research Note No. RN/3219/WAL.AWP, Crowthorne.

Lippomann, R. and Gudehus, G. (1989) Dowelled clay slopes: recent examples. *Proc. 12th International Conf. on Soil Mechanics and Foundation Engineering, Rio de Janeiro*, Vol. 2, pp. 1269–71.

Littlejohn, G. S. (1970) Soil anchors, in *Ground Engineering* (ed. J. S. Davis), Thomas Telford Press, London, pp. 33–45.

Littlejohn, G. S. (1982) Design of cement based grouts, in *Grouting in Geotechnical Engineering, Speciality Conf., New Orleans* (ed. W. H. Baker), American Society Civil Engineers, New York, pp. 35–48.

Littlejohn, G. S. (1985) Chemical grouting. *Ground Engineering*, **18**: Part 1, No. 2, pp. 13–18; Part 2, No. 3, pp. 23–8; Part 3, No. 4, pp. 29–34.

Littlejohn, G. S. (1990) Ground anchorage practice, in *Design and Performance of Earth Retaining Structures* (eds P. D. Lambe and L. A. Hansen), American Society Civil Engineers, Geotechnical Special Publication No. 25, pp. 692–733.

Litvinov, I. M. (1960) Stabilization of settling and weak clayey soils. *Highway Research Board Special Report No. 60*, Washington, D.C., pp. 94–112.

Litvinov, I. M. (1973) Deep compaction of soils with the aim of considerably increasing their bearing capacity. *Proc. 8th International Conf. on Soil Mechanics and Foundation Engineering, Moscow*, Vol. 3, pp. 392–4.

Lizzi, F. (1977) Practical engineering in structurally complex formations. The in situ reinforced earth. *Proc. International Symp. on Geotechnics in Complex Formations, Capri*, pp. 327–33.

Lo, K. Y., Inculet, I. I. and Ho, K. S. (1991) Electro-osmotic strengthening of soft sensitive clay. *Canadian Geotechnical Journal*, **28**, 62–73.

Lovell, C. W. (1957) *Temperature effects on phase composition and strength of partially frozen ground*. Highway Research Board, Bulletin No. 168, Washington, D.C.

Madu, R. M. (1977) An investigation into the geotechnical properties of some laterites of eastern Nigeria. *Engineering Geology*, **11**, 101–25.

Manfakh, G. R. (1989) Soil reinforcement – a tale of three walls. *Proc. 12th International Conf. on Soil Mechanics and Foundation Engineering, Rio de Janeiro*, Vol. 2, pp. 1285–8.

Mansur, C. I. and Kaufman, R. F. (1962) Dewatering, in *Foundation Engineering* (ed. G. A. Leonards), McGraw-Hill, New York, pp. 241–350.

McGown, A. (1971) The classification for engineering purposes of tills from moraines and associated landforms. *Quarterly Journal of Engineering Geology*, **4**, 115–30.

McGown, A. (1976) The properties and uses of permeable fabric membranes. *Proc. Residential Workshop on Materials and Methods for Low Cost Road, Rail and Reclamation Work, Leura, Australia*, pp. 663–709.

McGown, A. and Hughes, F. H. (1981) Practical aspects of the design and installation of deep vertical drains. *Geotechnique*, **31**, 3–18.

McGown, A., Kabir, M. H. and Murray, R. T. (1978) Compressibility and hydraulic conductivity of geotextiles. *Proc. 1st International Conf. on the Use of Fabrics in Geotechnics, Paris*, Vol. 2, pp. 167–72.

McGown, A., Andrawes, K. Z. and Kabir, M. H. (1982) Load extension properties of geotextiles confined in soil. *Proc. 2nd International Conf. on Geotextiles, Las Vegas*, Vol. 3, pp. 793–8.

McGown, A., Andrawes, K. Z., Hytiris, N. and Mercer, F. B. (1985) Soil strengthening using randomly distributed mesh elements. *Proc. 11th International Conf. on Soil Mechanics and Foundation Engineering, San Francisco*, Vol. 3, pp. 1735–8.

McKeand, E. and Sissons, C. R. (1978) Textile reinforcements – characteristic properties and their measurements. *Ground Engineering*, **11** (5), 13–16.

McKittrick, D. P. (1979) Reinforced earth: application of theory and research to practice. *Ground Engineering*, **12** (1), 19–31.

Menard, L. and Broise, Y. (1975) Theoretical and practical aspects of dynamic consolidation. *Geotechnique*, **15**, 3–18.

Mettier, K. (1985) Ground freezing for the construction of the Milchbruch tunnel in Zurich, Switzerland. *Proc. 4th International Symp. on Ground Freezing, Sapporo* (eds S. Kinosita and M. Fukuda), Balkema, Rotterdam, Vol. 1, pp. 263–70.

Meyerhof, G. G. (1951) Building on fills with special reference to settlement of a large factory. *Structural Engineer*, **29** (11), 297–305.

Miller, E. (1988) The eductor dewatering system. *Ground Engineering*, **21** (6), 29–34.

Millet, R. A. and Perez, J. Y. (1981) Current USA practice: slurry wall specifications. *Proc. American Society Civil Engineers, Journal Geotechnical Engineering Division*, **107** (GT8), 1041–56.

Minkov, M., Evstatiev, D. and Donchev, P. (1980) Dynamic compaction of loess. *Proc. International Conf. on Compaction, ENPC-LCPC, Paris*, pp. 345–9.

Mitchell, J. K. (1970) In place treatment of foundation soils. *Proc. American Society Civil Engineers, Journal Soil Mechanics and Foundations Division*, **96** (SM1), 73–110.

Mitchell, J. K. (1981) Soil improvement – State-of-the-art report. *Proc. 10th International Conf. on Soil Mechanics and Foundation Engineering, Stockholm*, Vol. 3, pp. 509–65.

Mitchell, J. K. and Wan, T. Y. (1977) Electro-osmotic consolidation – its effect on soft soils. *Proc. 9th International Conf. on Soil Mechanics and Foundation Engineering, Tokyo*, Vol. 1, pp. 219–24.

Mitchell, J. K. and Villet, W. C. B. (1987) *Reinforcement of Earth Slopes and Embankments*, National Cooperative Highway Research Program, Report 290, Transportation Research Board, National Research Council, Washington, D.C.

Mitchell, J. K. and Christopher, B. R. (1990) North American practice in reinforced soil systems, in *Design and Performance of Earth Retaining Structure* (eds P. C. Lambe and L. A. Hansen), American Society Civil Engineers, Geotechnical Special Publication No. 25, New York, pp. 322–46.

Moh, Z. C. and Woo, S. M. (1987) Preconsolidation of Bangkok Clay by nondisplacement sand drains and surcharge. *Proc. 9th South East Asian Geotechnical Conf., Bangkok*, pp. 8/171–8/174.

Moorhouse, D. C. and Baker, G. L. (1969) Sand densification by heavy vibratory compactor. *Proc. American Society Civil Engineers, Journal Soil Mechanics and Foundations Div.*, **95**, 985–94.

Nonveiller, E. (1989) *Grouting Theory and Practice*, Elsevier, Amsterdam.

North-Lewis, J. P. and Lyons, G. H. A. (1975) Continguous bored piles. *Proc. Conf. on Diaphragm Walls and Anchorages, London*, Institution Civil Engineers, pp. 189–94.

Nussbaum, P. J. and Colley, B. E. (1971) *Dam Construction and Facing with Soil-Cement*, Research and Development Bulletin No. RDOIOW, Portland Cement Association, Skokie, 14p.

O'Rourke, T. D. and Jones, C. J. F. P. (1990) Overview of earth retention systems: 1970–1990, in *Design and Performance of Earth Retaining Structures* (eds P. C. Lambe and L. A. Hansen), American Society Civil Engineers, Geotechnical Special Publication No. 25, pp. 22–51.

Parsons, A. W. (1987) Shallow compaction, in *Ground Engineer's Reference Book* (ed. F. G. Bell), Butterworths, London, pp. 37/3–37/17.

Pedley, M. J., Jewell, R. A. and Milligan, G. W. E. (1990) A large scale experimental study of soil-reinforced interaction. *Ground Engineering*, **20**: Part 1, No. 5, pp. 44–50; Part 2, No. 6, pp. 45–9.

Perry, W. (1963) Electro-osmosis dewaters large foundation excavation. *Construction Method*, **45** (9), 116–19.

Placzek, D. (1989) Methods for the calculation of settlements due to groundwater lowering. *Proc. 12th International Conf. on Soil Mechanics and Foundation Engineering, Rio de Janeiro*, Vol. 3, pp 1813–18.

Powers, J. P. (1981) *Construction Dewatering: A Guide to Theory and Practice*, Wiley Interscience, New York.

Powrie, W. and Roberts, T. O. L. (1990) Field trials of an ejector well dewatering system at Conwy, North Wales. *Quarterly Journal of Engineering Geology*, **23**, 169–85.

Prugh, B. J. (1960) New tools and techniques for dewatering. *Proc. American Society Civil Engineers, Journal Construction Division*, **80** (CO1), 11–25.

Prugh, B. J. (1963) Densification of soils by explosive vibration. *Proc. American Society Civil Engineers. Journal Construction Division*, **83** (CO1), 79–100.

Raabe, E. W. and Esters, K. (1990) Soil fracturing techniques for terminating settlements and restoring levels of buildings and structures. *Ground Engineering*, **23** (3), 33–45.

Raffle, J. F. and Greenwood, D. A. (1961) The relationship between the rheological characteristics of grouts and their capacity to permeate soils. *Proc. 5th International Conf. on Soil Mechanics and Foundation Engineering, Paris*, Vol. 2, pp. 789–93.

Rankilor, P. R. (1981) *Membranes in Ground Engineering*, Wiley, Chichester.

Rathmayer, H. G. and Saari, K. H. O. (eds) (1983) Improvement of ground. *Proc. 8th European Conf. on Soil Mechanics and Foundation Engineering, Helsinki*, Balkema, Rotterdam.

Ressi, A. and Cavalli, N. (1984) Bentonite slurry trenches. *Engineering Geology*, **21**, 333–9.

Richart, F. E. (1959) A review of theories for sand drains. *Proc. American Society Civil Engineers, Journal Soil Mechanics and Foundations Division*, **83** (SM3), 1301–38.

Robinson, K. E. and Eivemark, M. M. (1985) Soil improvement using wick drains and preloading. *Proc. 11th International Conf. on Soil Mechanics and Foundation Engineering, San Francisco*, Vol. 3, pp. 1739–44.

Rogers, C. D. F. (1991) Slope stabilization using lime. *Proc. International Conf. on Slope Stability Engineering, Developments and Applications, Isle of Wight*, Thomas Telford Press, London, pp. 395–402.

Rowe, P. W. (1964) The calculation of the consolidation rates of laminated, varved or layered clays with particular reference to sand drains. *Geotechnique*, **14**, 321–40.

Rowe, R. K., MacLean, M. D. and Barsvary, A. K. (1984) The observed behaviour of a geotextile-reinforced embankment constructed on peat. *Canadian Geotechnical Journal*, **21**, 289–304.

Ruffles, N. J. (1965) Derwent Reservoir. *Journal Institution Water Engineers*, **19**, 361–76.

Saito, A. (1977) Characteristics of penetration resistance of a reclaimed sandy deposit and their change through vibratory compaction. *Soils and Foundations*, **17**, 32–43.

Saitoh, S., Suzuki, Y. and Shirai, K. (1985) Hardening of soil improved by deep mixing method. *Proc. 11th International Conf. on Soil Mechanics and Foundation Engineering, San Francisco*, Vol. 3, pp. 1745–8.

Samol, H. and Priebel, H. (1985) Soil fracturing – an injection method for ground movement. *Proc. 11th International Conf. on Soil Mechanics and Foundation Engineering, San Francisco*, Vol. 3, pp. 1753–6.

Sanger, F. J. (1968) Ground freezing in construction. *Proc. American Society Civil Engineers, Journal Soil Mechanics and Foundations Division*, **94** (SM1), 131–58.

Sanger, F. J. and Sayles, F. H. (1979) Thermal and rheological computations for artificially frozen ground construction. *Proc. 1st Symp. on Ground Freezing, Bochum* (ed. H. L. Jessberger), Elsevier, Amsterdam, pp. 311–37; (also *Engineering Geology*, **13**).

Sayles, F. H. (1989) State-of-the-art: mechanical properties of frozen soil. *Proc. 5th International Symp. on Ground Freezing* (eds R. H. Jones and J. T. Holden), Balkema, Rotterdam, Vol. 1, pp. 143–66.

Sasaki, H. (1985) Effectiveness and applicability of the methods of foundation improvement for embankments over peat deposits, in *Recent Developments in Ground Improvement Techniques* (eds A. S. Balasubramanian, S. Chandra, D. T. Bergado *et al.*), Balkema, Rotterdam, pp. 543–62.

Schlosser, F. (1987) Reinforced Earth, in *Ground Engineer's Reference Book* (ed. F. G. Bell), Butterworths, London, pp. 35/1–35/15.

Schlosser, F. (1990) Mechanically stabilized earth retaining structures in Europe, in *Design and Performance of Earth Retaining Structures* (eds P. C. Lambe and L. A. Hansen), American Society Civil Engineers, Geotechnical Special Publication No. 25, pp. 347–78.

Schlosser, F. and Guilloux, A. (1981) Le frottement dans le reinforcement des sols. *Revue Francaise de Geotechnique*, **18**, 66–71.

Schweitzer, F. (1989) Strength and permeability of single-phase diaphragm walls. *Proc. 12th International Conf. on Soil Mechanics and Foundation Engineering, Rio de Janiero*, Vol. 3, pp. 1515–18.

Sherard, J. L., Dunnigan, L. P. and Dekker, R. S. (1976) Identification and nature of dispersive soils. *Proc. American Society of Civil Engineers, Journal Geotechnical Engineering Division*, **102** (GT4), 287–301.

Sherard, J. L., Dunnigan, L. P. and Talbot, J. R. (1984a) Basic properties of sand and gravel filters. *Proc. American Society Civil Engineers, Journal Geotechnical Engineering Division*, **110** (GT6), 684–700.

Sherard, J. L., Dunnigan, L. P. and Talbot, J. R. (1984b) Filters for silts and clays. *Proc. American Society Civil Engineers, Journal Geotechnical Engineering Division*, **110** (GT6), 701–18.

Sherwood, F. T. (1957) The stabilization with cement of weathered and sulphate-bearing clays. *Geotechnique*, **7**, 179–91.

Shibazaki, M. and Ohta, S. (1982) A unique underpinning of soil solidification utilizing super-high pressure liquid jet. *Grouting in Geotechnical Engineering, Speciality Conf., New Orleans* (ed. W. H. Baker), American Society Civil Engineers, New York, pp. 680–93.

Shuster, J. A. (1972) Controlled freezing for temporary ground support. *Proc. 1st North American Rapid Excavation and Tunneling Conf., Chicago*, Vol. 2, pp. 863–94.

Sliwinski, Z. and Fleming, W. G. K. (1975) Practical considerations affecting the construction of diaphragm walls. *Proc. Conf. on Diaphragm Walls and Anchorages*, Institution of Civil Engineers, London, pp. 1–10.

Skempton, A. W. (1964) The long-term stability of clay slopes. *Geotechnique*, **14**, 77–101.

Skempton, A. W., Schuster, R. L. and Petley, D. J. (1969) Joints and fissures in the London Clay at Wraysbury and Edgeware. *Geotechnique*, **19**, 205–17.

Skipp, B. O. and Renner, L. (1963) The improvement of the mechanical properties of sands. *Proc. Symp. on Grouts and Drilling Muds in Engineering Practice*, Butterworths, London, pp. 29–35.

Somerville, S. H. (1986) *Control of Groundwater for Temporary Works*, Construction Industry Research and Information Associations, Report 113, London.

Sowers, G. F. (1973) Settlement of waste disposal fills. *Proc. 8th International Conf. on Soil Mechanics and Foundation Engineering, Moscow*, Vol. 2, pp. 207–12.

Sridharan, A., Srinivisa Murthy, B. R., Bindumadhava and Revanasiddappa, K. (1991) Technique for using fine grained soil in reinforced earth. *Proc. American Society Civil Engineers, J. Geotechnical Engineering Division*, **117** (8), 1174–90.

Stilley, A. N. (1982) Compaction grouting for foundation stabilization, in *Grouting in Geotechnical Engineering, Speciality Conf., New Orleans* (ed. W. H. Baker), American Society Civil Engineers, New York, pp. 923–37.

Stocker, M. F. and Zwicker, C. A. (1987) Soil improvement by high pressure injection for the foundation of a highrise building in Kuala Lampur. *Proc. 9th South East Asian Geotechnical Conf., Bangkok*, pp. 8/113–8/138.

Talbot, J. R. (1991) Discussion: Improved filter criterion for cohesionless soils. *Proc. American Society Civil Engineers, Journal Geotechnical Engineering Division*, **117**, 1633–4.

Tan, D. Y. and Clough, G. W. (1980) Ground control for shallow tunnels by soil grouting. *Proc. American Society of Civil Engineers, Journal Geotechnical Engineering Division*, **106** (GT9), 1037–57.

Tang, Y. and Gao, Z. (1989) Experimental study and application of vacuum preloading for consolidating soft soil foundation. *Proc. 12th International Conf. on Soil Mechanics and Foundation Engineering, Rio de Janiero*, Vol. 2, pp. 1423–6.

Tausch, N. (1985) A special grouting method to construct horizontal membranes. *Proc. International Symp. on Recent Developments in Ground Improvement Techniques, Bangkok* (eds A. S. Balasubramanian, S. Chandra, D. T. Bergado, Younger, J. S. and Prinzl, F.), Balkema, Rotterdam, pp. 351–62.

Temporal, J., Craig, A. H., Harris, D. H. and Brady, K. C. (1989) The use of locally available fills for reinforced and anchored earth. *Proc. 12th International Conf. on Soil Mechanics and Foundation Engineering, Rio de Janeiro*, Vol. 2, pp. 1315–20.

Terzaghi, K. (1925) *Erdbaumechanik auf Bodenphysikalischer Grundlage*, Deuticke, Wien.

Terzaghi, K. (1943) *Soil Mechanics*, Wiley, New York.

Thorburn, S. (1975) Building structures supported by stabilized ground. *Geotechnique*, **25**, 83–94.

Tomlinson, M. J. (1986) *Foundation Design and Construction*, 5th edn, Longmans, London.

Toombes, A. F. (1969) *The Performance of an Aveling Barford VP 7.7 Mg Self-propelled Vibrating Roller in the Compaction of Soil*, Transport and Road Research Laboratory, Report LR257, Crowthorne.

Tsytovich, N. A., Abelev, M. Y. and Takhirov, I. G. (1971) Compacting saturated loess soils by means of lime piles. *Proc. 4th European Conf. on Soil Mechanics and Foundation Engineering, Budapest*, pp. 837–42.

Van der Merwe, C. J. and Horak, E. (1989) Evaluation of soil/geotextile compatibility. *Proc. 12th International Conf. on Soil Mechanics and Foundation Engineering, Rio de Janeiro*, Vol. 3, pp. 1671–7.

Further reading

General

American Society of Civil Engineers (1987) *Soil Improvement: A Ten Year Update*, Geotechnical Engineering Division, Geotechnical Special Publication No. 12, New York.

Balasubramaniam, A. S., Chandra, S., Bergado, D. T., Younger, J. S. and Prinzl, F. (eds) (1985) *Recent Developments in Ground Improvement Techniques*, Balkema, Rotterdam.

Bell, F. G. (1975) *Methods of Treatment of Unstable Ground*, Newnes–Butterworths, London.

Bell, F. G. (ed.) (1987) *Ground Engineer's Reference Book. Part 3, Treatment of the Ground*, Butterworths, London.

British Standards Institution (1981) *Code of Practice on Earthworks*, BS 6031, London.

David, J. S. (ed.) (1970) *Ground Engineering*, Institution of Civil Engineers, Thomas Telford Press, London.

Greenwood, D. A. and Thomson, G. H. (1984) *Ground Stabilization: Deep Compaction and Grouting*, Institution of Civil Engineers, Thomas Telford Press, London.

Hausmann, M. R. (1990) *Engineering Principles of Ground Modification*, McGraw-Hill, New York.

Institution of Civil Engineers, (1983) *Advances in Piling and Ground Treatment*, Thomas Telford Press, London.

Rathmayer, H. G. and Saari, K. H. O. (eds) (1983) *Improvement of ground*, Proc. 8th European Conf. on Soil Mechanics and Foundation Engineering, Helsinki, Balkema, Rotterdam (3 volumes).

Van Impe, W. F. (1989) *Soil Improvement Techniques and Their Evolution*, Balkema, Rotterdam.

Exclusion techniques

Hajnal, I., Martin, J. and Regele, Z. (1984) *Construction of Diaphragm Walls*, Wiley Interscience, New York.

Institution of Civil Engineers, (1975) *Diaphragm Walls and Anchorages*, Thomas Telford Press, London.

Xanthakos, P. P. (1979) *Slurry Walls*, McGraw-Hill, New York.

Ground freezing

Frivik, P. E., Janbu, H., Saetersdal, R. and Finerud, L. I. (eds) (1981) *Ground Freezing: Developments in Geotechnical Engineering*, No. 28, Elsevier, Amsterdam (also *Engineering Geology*, **18**.)
Jessberger, H. (ed.) (1979) *Ground Freezing: Developments in Geotechnical Engineering*, No. 26, Elsevier, Amsterdam (also *Engineering Geology*, **13**.)
Chamberlain, E. G. (ed.) (1982) *Ground Freezing*, Cold Regions, Research and Engineering Laboratory, Special Report 82–16, Hanover, New Hampshire.
Kinosita, S. and Fukuda, M. (eds) (1985) *Ground Freezing*, Balkema, Rotterdam.
Jones, R. H. and Holden, T. (eds) (1989) *Ground Freezing '88*, Balkema, Rotterdam (2 volumes).
Xiang, Y. and Changsheng, W. (1991) *Ground Freezing '91*, Balkema, Rotterdam (2 volumes).

Drainage and dewatering

Cedergren, H. (1986) *Seepage, Drainage and Flow Nets*, 3rd edn, Wiley, New York.
Holtz, R. D., Jamiolkowski, M., Lancelotta, R. and Pedroni, S. (1992) *Performance of Prefabricated Band Shaped Drains*, Construction Industry Research and Information Association, London.
Institution of Civil Engineers, (1982) *Vertical Drains*, Thomas Telford Press, London.
Powers, J. P. (1981) *Construction Dewatering: A Guide to Theory and Practice*, Wiley Interscience, New York.
Somerville, S. H. (1988) *Control of Groundwater for Temporary Works*, Report 113, Construction Industry Research and Information Association, London.

Compaction techniques

Institution of Civil Engineers, (1976) *Ground Improvement by Deep Compaction*, Thomas Telford Press, London.
Institution of Civil Engineers, (1978) *Clay Fills*, Thomas Telford Press, London.
Stamatopoulos, A. C. and Kotzias, P. C. (1985) *Soil Improvement by Preloading*, Wiley Interscience, New York.

Soil reinforcement

American Society of Civil Engineers, (1978) *Symposium on Earth Reinforcement, Pittsburgh*, New York.
British Standards Institution, (1991) *Code of Practice for Strengthened/Reinforced Soils and Other Fills*, BS 8006, London.
Christopher B. R., Gill, S. A., Giroud, J. P., Juran, I., Schlosser, F., Mitchell, J. K. and Donnicliff, J., (1991) *Reinforced Soil Structures. Vol. 1: Design and Construction Guidelines*, Report No. FHWA-RD-89-043, Federal Highways Administration, Washington, D.C.
Hanna, T. H. (1982) *Foundations in Tension: Ground Anchors*, Trans. Tech. Publications, Clausthal.
Hobst, L. and Zajic, J. (1983) *Anchoring in Rock and Soil*, 2nd edn, Elsevier, Amsterdam.
Ingold, T. S. (1982) *Reinforced Earth*, Institution of Civil Engineers, Thomas Telford Press, London.
Jones, C. J. F. P. (1985) *Reinforced Earth*, Butterworths, London.

Lambe, P. C. and Hansen, L. A. (eds) (1990) *Design and Performance of Earth Retaining Structures*, American Society of Civil Engineers, Geotechnical Special Publication No. 25, New York.
McGown, A., Yeo, K. C. and Andrawes, K. Z. (eds) (1991) *Performance of Reinforced Soil Structures*, Thomas Telford Press, London.
Mitchell, J. K. and Villet, C. B. (eds) (1987) *Reinforcement of Earth Slopes and Embankments*, National Cooperative Highway Research Program Report 290, Transportation Research Board, Washington, D.C.
Shercliff, D. A. (ed.) (1989) *Reinforced Embankments: Theory and Practice*, Thomas Telford Press, London.
Technical Memorandum BE3/78, (1987) *Reinforced Earth Retaining Walls and Bridge Abutments for Embankments*, Department of Transport, HMSO, London.
Yamanouchi, T., Miara, N. and Ochiai, H. (eds) (1988) *Theory and Practice of Earth Reinforcement*, Balkema, Rotterdam.

Geosynthetics

Anon. (1984) *A Geotextile Design Guide*, Low Bros, Dundee.
Den Hoedt, G. (ed.) (1990) *Geotextiles, Geomembranes and Related Products*, Balkema, Rotterdam (3 volumes).
Holtz, R. D. (ed.) (1988) *Geosynthetics for Soil Improvement*, American Society of Civil Engineers, Geotechnical Special Publication No. 18. New York.
Ingold, T. S. and Miller, K. S. (1988) *Geotextiles Handbook*, Thomas Telford Press, London.
Institution of Civil Engineers, (1984) *Proc. of Symposium on Polymer Grid Reinforcements*, Thomas Telford Press, London.
Jarrett, P. M. and McGown, A. (eds) (1987) *The Application of Polymeric Reinforcement in Soil Retaining Structures*, NATO Advanced Study Institute Series, Kluwer, Holland.
John N. W. M. (1987) *Geotextiles*, Blackie, Glasgow.
Koener, R. M. (1986) *Designing with Geotextiles*, Prentice-Hall, Englewood Cliffs, New Jersey.
Koerner, R. M. and Welsh, J. P. (1980) *Construction and Geotechnical Engineering using Synthetic Fabrics*, Wiley, New York.
Rankilor, P. R. (1981) *Membranes in Ground Engineering*, Wiley, Chichester.
Van Zanten, R. V. (1986) *Geotextiles and Geomembranes in Civil Engineering*, Wiley, New York.

Grouting

Baker, W. H. (ed.) (1982) *Grouting in Geotechnical Engineering. Speciality Conference, New Orleans*, American Society of Civil Engineers, New York.
Bowen, R. (1981) *Grouting in Engineering Practice*, 2nd edn, Applied Science, London.
Cambefort, H. (1967) *Injection des Sols*, Eyrolles, Paris.
Engineering Manual No. 1110-2-3506, (1984) *Grouting Technology*, US Corps of Engineers, Washington, D.C.
Houlsby, A. C. (1990) *Construction and Design of Cement Grouting*, Wiley Interscience, New York.
Karol, R. H. (1983) *Chemical Grouting*, Marcel Dekker, New York.
Nonveiller, E. (1989) *Grouting Theory and Practice*, Elsevier, Amsterdam.

Verfel, J. (1989) *Rock Grouting and Diaphragm Wall Construction*, Elsevier, Amsterdam.

Soil stabilization

Imperial Chemical Industries, (1986) *Lime Stabilization Manual*, Buxton, Derbyshire.

Ingles, O. G. and Metcalf, J. B. (1977) *Soil Stabilization*, 2nd edn, Butterworths, Sydney.

National Lime Association, (1982) *Lime Stabilization Construction Manual*, 7th edn, Washington, D.C.

Winterhorn, H. F. and Schanid, W. E. (1971) *Soil Stabilization Parameters*, Technical Report No. AFWL-TR-70-35, Kirtland Air Force Base, New Mexico.

Site investigation

British Standards Institution, (1981) *Code of Practice on Site Investigation*, BS 5930, London.

Clayton, C. R., Simons, N. E. and Matthews, M. C. (1982) *Site Investigation*, Granada, London.

Hawkins, A. B. (ed) (1986) *Site Investigation Practice: Assessing BS 5930*, Engineering Geology Special Publication No. 2, The Geological Society, London.

Hvorslev, M. J. (1951) *Subsurface Exploration and Sampling of Soils for Civil Engineering Purposes*, Waterways Experimental Station, Vicksburg, Mississippi.

Weltman, A. J. and Head, J. M. (1983) *Site Investigation Manual*, Construction Industry Research and Information Association, Special Publication 25, London.

Soil types

Bell, F. G. (1992) *Engineering Properties of Soils and Rocks*, 3rd edn, Butterworth–Heinemann, Oxford.

Geotechnical Control Office, (1986) *Guide to Rock and Soil Descriptions, Geoguide 3*, Government Publication Centre, Hong Kong.

Gillott, J. E. (1987) *Clay in Engineering Geology*, 2nd edn, Elsevier, Amsterdam.

Mitchell, J. K. (1976) *Fundamentals of Soil Behavior*, Wiley, New York.

Young, R. N. and Warkentin, B. P. (1975) *Soil Properties and Behaviour*, Elsevier, Amsterdam.

Index

Page numbers in **bold** type refer to illustrations.